Civil
Drafting
Technology

CIVIL DRAFTING TECHNOLOGY

David A. Madsen

Department Chairperson, Drafting Technology
Clackamas Community College
Oregon City, Oregon

Terence M. Shumaker

Instructor, Drafting Technology
Clackamas Community College
Oregon City, Oregon

PRENTICE-HALL, INC. Englewood Cliffs, New Jersey 07632

Library of Congress Cataloging in Publication Data

Madsen, David A.
    Civil drafting technology.

    Includes index.
    1. Mechanical drawing.   I. Shumaker, Terence M.
II. Title.
T353.M196 1983   624'.022'1   82-12252
ISBN 0-13-134890-6

Editorial/production supervision and
    interior design by *Shari Ingerman*
Page layout by *Mary Greey*
Cover design by *Photo Plus Art*
Manufacturing buyer: *Tony Caruso*

Printed in the United States of America

10

ISBN 0-13-134890-6

PRENTICE-HALL INTERNATIONAL, INC., *London*
PRENTICE-HALL OF AUSTRALIA PTY. LIMITED, *Sydney*
EDITORA PRENTICE-HALL DO BRASIL, LTDA., *Rio de Janeiro*
PRENTICE-HALL CANADA INC., *Toronto*
PRENTICE-HALL OF INDIA PRIVATE LIMITED, *New Delhi*
PRENTICE-HALL OF JAPAN, INC., *Tokyo*
PRENTICE-HALL OF SOUTHEAST ASIA PTE. LTD., *Singapore*
WHITEHALL BOOKS LIMITED, *Wellington, New Zealand*

Contents

Preface

This text is intended to be a comprehensive instructional package in the area of civil drafting. The authors have used materials that have been tested in the classroom for several years plus input and ideas from industry, specifically civil engineering companies. Our aim is to provide the student or employee with a well-rounded view of the civil drafting field and the types of drawings and skills associated with that field.

The book is arranged in ten chapters, each dealing with a specific subject area. We feel that the arrangement lends itself well to a one-term course (ten to twelve weeks long) but contains enough information and problems to fit courses of varying length. Each chapter is followed by a test or tests composed of short essay questions, fill-ins, true/false, and sketching. Problems in the form of drawings, which close each chapter, can be completed in the text or on separate sheets of paper. The tests and problems enable the student to apply directly the information given in each chapter to realistic situations.

The field of civil drafting is one filled with variety and excitement. From surveying to construction, courthouse research to artistic interpretation, the opportunities offer many challenges. The authors have drawn on their experiences of surveying in the jungles of Georgia to building houses in the foothills of the Oregon Cascades; from designing wastewater piping to designing solar homes. And most important, we have drawn on our collective experiences in teaching the varied aspects of drafting at the community college level.

The use of this text in the prescribed manner will impart to the student a broad knowledge of civil drafting and a working knowledge of the basic components of mapping. With this knowledge and skill a variety of job opportunities will open to the student, and with those opportunities, we sincerely hope, a challenging career. Keep in mind that mapping requires accuracy, neatness, and an eye for creative and uncluttered layout.

Our thanks must be extended to the Oregon Department of Education for allowing us to draft from some of their illustrations.

We hope that your experiences with civil drafting and mapping are just as exciting as ours. Good luck.

David Madsen
Terence Shumaker

Civil
Drafting
Technology

1

Introduction

This chapter discusses maps in general and some of the different types in use today. Information about civil engineering companies, their map drafting requirements, and employment opportunities is also covered.

Topics to be discussed include:

- Characteristics of maps
- Types of maps
- Civil engineering companies
- Map requirements
- Schooling
- Cartography

MAPS IN GENERAL   Maps are defined as graphic representations of part of or the entire earth's surface drawn to scale on a plane surface. Constructed and natural features may be shown by lines, symbols, and colors. Maps have many different purposes depending on their intended usage. A map can accurately provide distances, locations, elevations, best routes, terrain features, and much more.

Some maps, such as aeronautical and nautical maps, are more commonly referred to as charts. This distinction is shown in the following discussion about types of maps.

Map Title Block and Legend

When you use or read a map the first place to look is the title block and legend. The information given here will tell immediately if you have the correct map. Other valuable information about map scales, symbols, compass direction, and special notes will also be given.

TYPES OF MAPS   Aeronautical Charts

Aeronautical charts are used as an aid to air travel. These charts indicate important features of land, such as mountains and outstanding landmarks. Commonly prepared in color and with relief-shading methods, aeronautical charts are a very descriptive representation of a portion of the earth's surface. Contour lines are often provided with 200- to 1000-ft. intervals. There is a comprehensive amount of information regarding air routes, airport locations, types of air traffic, radio aids to navigation, and maximum elevation of features. Take a look at Figure 1-1 and you can see all of the detail shown in an aeronautical chart.

Figure 1-1 A typical aeronautical chart. Reproduced by permission of the National Ocean Survey, (NOAA), U.S. Department of Commerce.

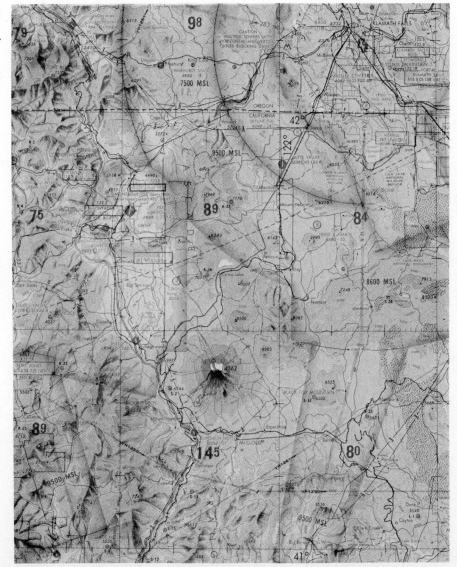

Cadastral Maps

Cadastral maps are large-scale maps that accurately show the features in a city or town. These types of maps are often used for city development, operation, and taxation. Figure 1-2 provides an example of a cadastral map.

Figure 1-2   A typical cadastral map. Reproduced by permission of the U.S. Geological Survey.

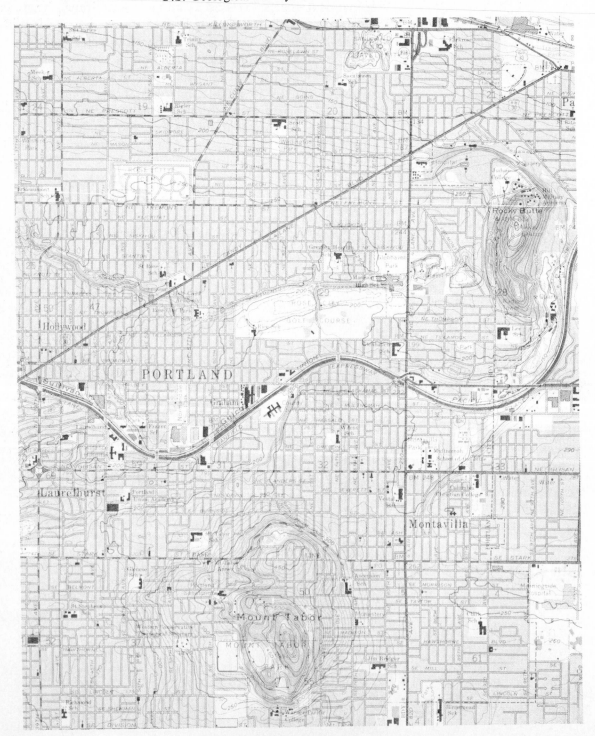

Engineering Maps

Construction projects of all kinds are detailed to show the complete layout in an engineering map. The information provided includes the location and dimensions of all structures, roads, parking areas, drainage ways, sewers, and other utilities. Elevations of features and contour lines are optional. See Figure 1-3 for an example.

Engineering maps may also include plats. Plats are carefully detailed maps of construction projects such as subdivisions showing building lots. These may also be plot or site plans, which are plats of an individual construction site. Figures 1-4 and 1-5 provide examples.

Geographical Maps

Geographical maps are usually prepared at a small scale. These maps commonly show large areas of the earth, depicting continents, countries, cities, rivers, and other important features. A map of the world or maps of individual countries or states are considered geographical maps. Figure 1-6 represents a typical geographical map.

Figure 1-3 Typical engineering site plans.

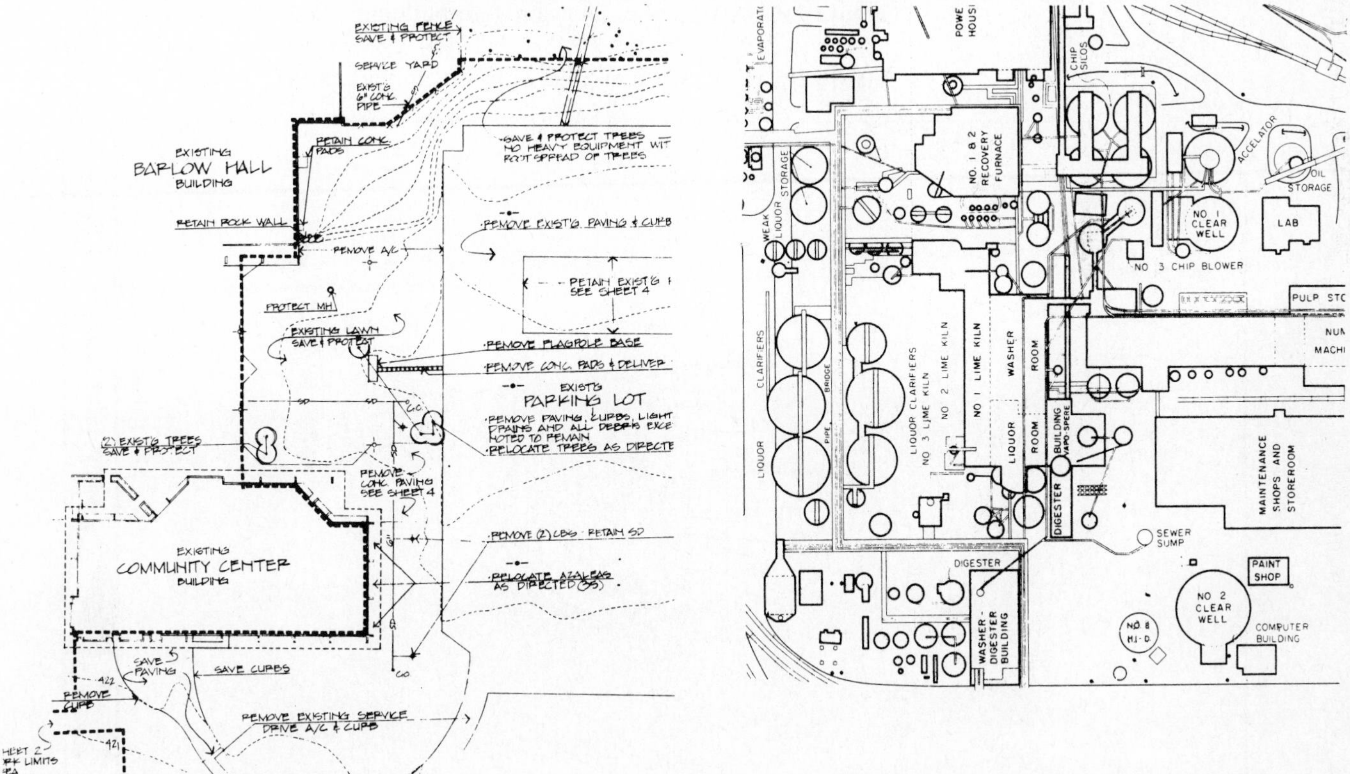

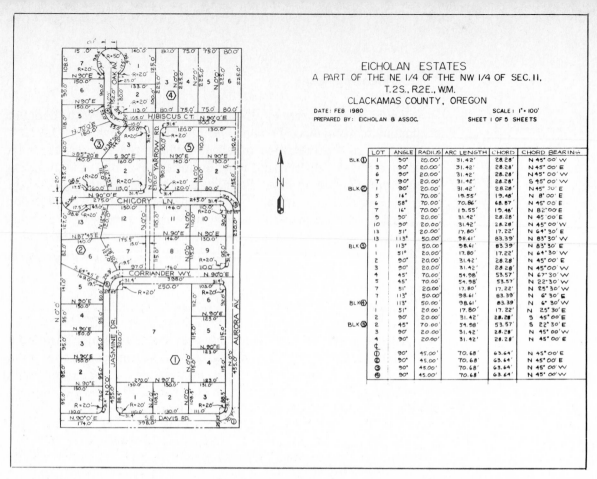

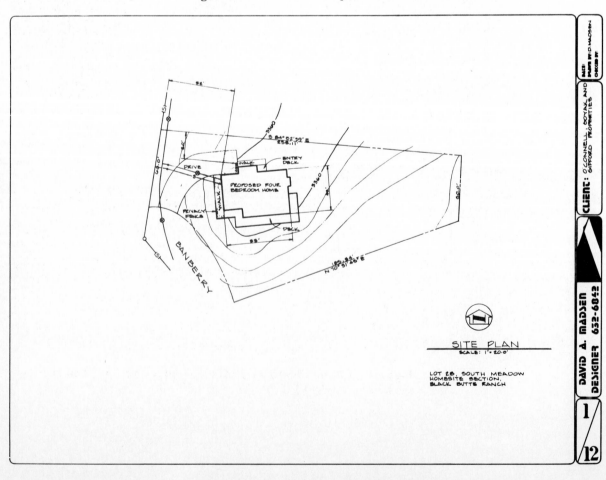

Figure 1-4 A typical subdivision with building lots.

Figure 1-5 An individual plat with residential construction site.

Figure 1-6 A typical geographical map. Reproduced by permission of the U.S. Geological Survey.

7

Hydrologic Maps

Hydrologic maps accurately show the hydrographic boundaries of major river basins. In the United States these maps are prepared by the U.S. Geological Survey in cooperation with the U.S. Water Resources Council. Hydrologic maps, which are used for water and land resource planning, are published at a scale of 1:500,000 (1 inch equals about 8 miles). The maps are printed in color and contain information on drainage, culture, and hydrographic boundaries. Figure 1-7 provides a sample hydrologic map.

Figure 1-7 A typical Hydrologic Map. Reproduced by permission of the U.S. Geological Survey.

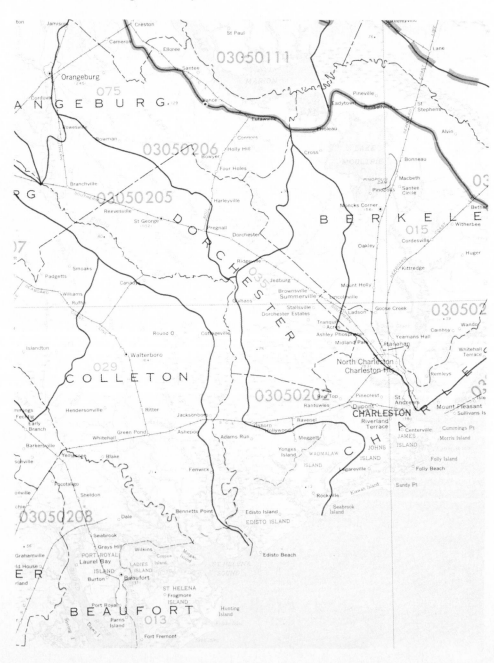

Military Maps

Military maps can be any map that contains information of military importance or serves a military use. A military map may be used by a soldier in the field and may have information about terrain, concealment, and cover. It may also be a map of a large geographical area that can be used for military planning.

Nautical Charts

Nautical charts are special maps used specifically as an aid to water navigation. These charts provide such information as water depths, clearances of bridges, and overhead cables. They also show navigation lanes, lighthouses, beacons, and buoys. Figure 1-8 shows a sample nautical chart.

Figure 1-8 A typical Nautical Chart. Reproduced by permission of the National Ocean Survey (NOAA), U.S. Department of Commerce.

Photogrammetric Maps

Aerial photographs are used to make photogrammetric maps. This process is the most widely used method of preparing maps. Aerial photos are taken at certain intervals and are controlled by stations on the earth's surface. These photos can be accurately scaled, and are easily read and transferred to paper using special stereoscopic instruments. Figure 1-9 shows an aerial photograph.

Compare the map in Figure 1-10 to the aerial photograph in Figure 1-9 and you can see how a map is created from an aerial photograph.

Figure 1-9   A typical Photogrammetric Map. Reproduced by permission of the U.S. Geological Survey.

Figure 1-10   A map created from an aerial photograph. Reproduced by permission of the U.S. Geological Survey.

Topographic Maps

Topographic maps accurately show the shape of the earth by the use of contour lines. Contour lines represent lines on the ground of equal elevation above sea level. The spacing of the contour lines is determined by the grade of the land. On very steep terrain the lines are close together, because changes in elevation come quickly. On terrain that slopes gradually the contour lines are farther apart, as it takes longer to reach a change in elevation. The contour lines are drawn at equal changes in elevation: for example, every 2, 5, or 10 ft. Usually, at least every fifth contour is broken along its length and the elevation is inserted. Contour lines are often drawn in brown.

11

Topographic maps show streams, lakes, and rivers in blue. Wood-land features are represented in green. Features that are constructed by people, such as buildings and roads, are often shown in black on a topographic map. Most of the types of maps previously discussed may often use the elements of a topographic map. Most construction projects involving shape, size, location, slope, or configuration of land can be aided by the use of topographic maps. There are probably thousands of varied uses for these maps. Figure 1-11 shows a topographic map.

Figure 1-11   A typical Topographic Map. Reproduced by permission of the U.S. Geological Survey.

CIVIL ENGINEERING COMPANIES

Civil engineering is concerned with the design of bridges, roads, dams, and canals. Civil engineering companies are located nationwide in most cities. Some of these companies specialize in certain aspects of the industry, whereas others are quite diversified. The following is a list of some of the tasks that civil engineering companies may take part in:

- Land planning and subdivision
- Transportation
- Flood control
- Irrigation and drainage
- Sewage and water treatment
- Municipal improvements
- Environmental studies
- Land and construction surveys
- Construction inspection
- Refuse disposal
- Mapmaking
- Power plants
- Hydrologic studies
- Foundation work and soil analysis
- Agribusiness

A complete directory of consulting engineers is available from the American Consulting Engineers Council.

Drafting salaries are usually competitive with other technologies. Working conditions vary but are usually excellent. Companies have a wide range of employee benefits. Check your local area regarding salary ranges and schooling requirements for entry-level drafters. Areas of the country differ in these concerns.

MAP REQUIREMENTS

Maps serve a multitude of purposes. Some maps may be used to show the construction site for a new home, while another map may show the geography of the world. Civil engineering companies primarily prepare maps that fall into the first category: construction site plans and maps relating to the civil projects previously described. The chapters covered in this workbook provide you with the basic information to continue a more in-depth study or on-the-job training.

The materials in most common use in civil drafting are vellum and pencil. Vellum provides a readily reproducible original at minimal cost. Companies that require a more durable original may provide their drafting technicians with Mylar. Mylar reproduces better than vellum but is more costly. When Mylar is used, the drafting technician uses polyester lead or ink to create the map. A more detailed description of drafting materials can be found in most basic drafting texts.

SCHOOLING   Technical schools and community colleges throughout the United States have drafting programs. Schools may provide a specific drafting education in mechanical, architectural, civil, piping, structural, technical illustration, sheet metal, electrical, or computer drafting. Other drafting programs may provide their students with a more general curriculum that may have courses in each of several of these areas. Often, the school focuses on the industrial needs of the immediate area. The best thing to do is to identify the school program that will best serve your specific goals. Civil drafting is offered in many technical schools and community colleges.

If you are interested in civil drafting as a career, your schooling should include the development of some of these skills:

- Drafting skills, line work, lettering, neatness, and the use of equipment
- Use of bearings and azimuths
- Use of the engineer's scale
- Scale conversion
- Drawing contour lines, and converting field notes
- Use of mapping symbols
- Preparation of a plat and interpretation of legal descriptions
- Development of plans and profiles
- Layout of highways, centerlines, curves, and delta angles
- Drawing cuts and fills
- Basic use and knowledge of surveying equipment
- Math through basic trigonometry

CARTOGRAPHY   Cartography is the art of making maps and charts. A cartographer is a highly skilled professional who designs and draws maps. Cartography is considered an art. The cartographer is a master in the use of a variety of graphic media, mechanical lettering methods, and artistic illustration.

Civil drafting and cartography are quite similar in that both professions deal with the making of maps. However, civil drafting is generally concerned with maps and plans for construction and other civil-related projects. Cartography requires that the technician use more graphic skills in the preparation of printed documents and maps. Often the job title "cartographer" requires four years of education, with emphasis in civil engineering, geography, navigation, optics, geodesy, or cartography.

TEST

Part I

Define the following terms. Use your best lettering technique.

1.1. Map GRAPHIC REPRESENTATION OF PART OF OR THE ENTIRE EARTH'S SURFACE DRAWN TO SCALE ON A PLANE SURFACE.

1.2. Contour lines REPRESENT LINES ON THE GROUND OF EQUAL ELEVATION ABOVE SEA LEVEL

1.3. Aerial photos TAKEN AT CERTAIN INTERVALS AND ARE CONTROLLED BY STATIONS ON THE EARTH'S SURFACE

1.4. Plats CAREFULLY DETAILED MAPS OF CONSTRUCTION PROJECTS SUCH AS SUBDIVISIONS SHOWING BUILDING LOTS.

1.5. Civil drafting DEALS WITH THE DESIGN OF BRIDGES, ROADS, DAMS AND CANALS

Part II

Multiple choice: Circle the response that best describes each statement.

1.1. Small-scale maps that commonly show large areas of earth, depicting continents and countries are called:
    a.   Aeronautical charts
    b.   Cadastral maps
    c.   Geographical maps
    d.   Nautical charts

15

1.2.   Maps that accurately show the shape of the earth by the use of contour lines are called:
   a.   Photogrammetric maps
   b.   Topographic maps
   c.   Engineering maps
   d.   Geographic maps

1.3.   Maps that accurately show the boundaries of major river basins are called:
   a.   Geographical maps
   b.   Nautical charts
   c.   Cadastral maps
   d.   Hydrologic maps

1.4.   Maps detailed to show the layout of a construction project are called:
   a.   Geographical maps
   b.   Cadastral maps
   c.   Engineering maps
   d.   Topographic maps

PROBLEMS

P1.1.   Below each line on Figure P1-1, draw five more lines exactly the same using your drafting pencil and a straightedge. Make your lines as dark and crisp as you would on a real drawing.

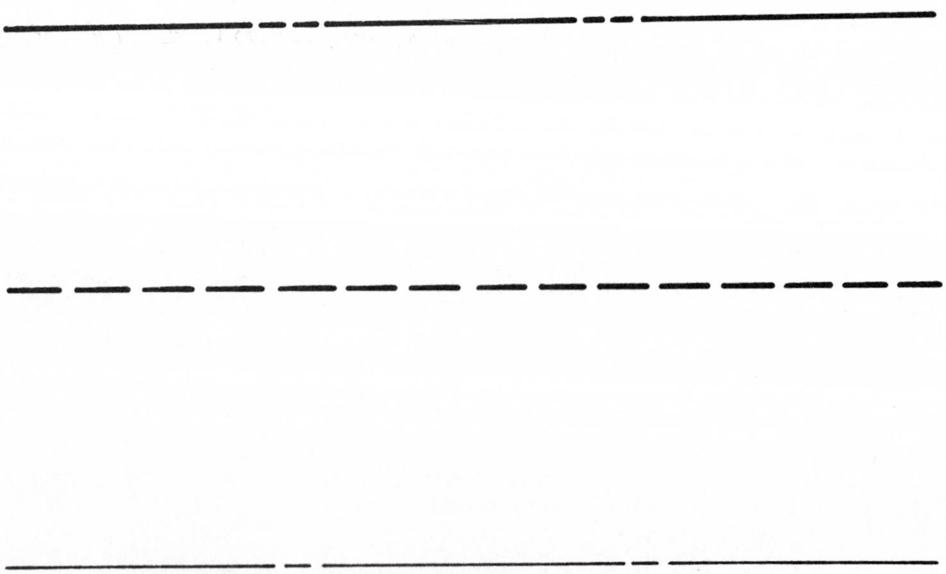

Figure P1-1

P1.2.   Using your best lettering skill, duplicate each sentence written on Figure P1-2. Make your own very light guidelines 1/8 in. apart to guide your lettering. Try to make your lettering the same as the example unless otherwise indicated by your instructor. Use a soft, slightly rounded pencil point.

THE QUALITY OF THE FREEHAND LETTERING GREATLY
AFFECTS THE APPEARANCE OF THE ENTIRE DRAWING.
MANY CIVIL DRAFTING TECHNICIANS USE FREEHAND
LETTERING IN PENCIL OR INK TO CREATE MAPS.
PROPER FREEHAND LETTERING IS DONE WITH A SOFT,
SLIGHTLY ROUNDED POINT DRAFTING PENCIL 2H, H, OR
F DEPENDING UPON THE INDIVIDUAL PRESSURE.   THE
LETTERING IS DONE BETWEEN VERY LIGHTLY DRAWN
GUIDE LINES.   THESE GUIDE LINES ARE DRAWN
PARALLEL SPACED EQUAL TO THE HEIGHTS OF THE
LETTERS.   GUIDE LINES HELP TO KEEP YOUR
LETTERING UNIFORM IN HEIGHT. LETTERING STYLES
MAY VARY BETWEEN COMPANIES, WHILE SOME
COMPANIES REQUIRE MECHANICAL LETTERING DEVICES.

Figure P1-2

2

Surveying Fundamentals

Maps are created for specific purposes and vary greatly from one type to another. A topographic map is used for a variety of things that you could not consider using a property plat for. And just as the usage of a map is different, so are the methods used to create them. This chapter examines some of those methods and the instruments used in gathering information from which maps are made.

TYPES OF SURVEYS

Land or Boundary Surveys

Most of us are familiar with this type of survey. If you own land, there probably exists a property plat of the plot at your county courthouse. This type of survey locates property corners and boundary lines. It is normally a *closed traverse* because the survey always returns to the *point of beginning* (POB) or another control point. An example of a land survey is shown in Figure 2-1.

Topographic Survey

Anyone who has ever worked with a contour map has seen the results of a topographic survey. The principal function of this survey is to locate elevations and features on the land, both natural or artificial (see Figure 2-2).

Geodetic Survey

This is a grand survey, often spanning nations, in which the curvature of the earth is a factor. Large areas are mapped by a process

Figure 2-1 Typical land survey of a subdivision (Courtesy OTAK & Associates, Inc.)

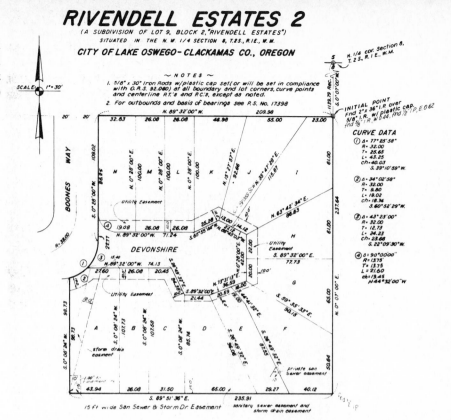

Figure 2-2 A topographic survey is used to compile the information needed to create this topographic map. (Courtesy U.S. Geological Survey)

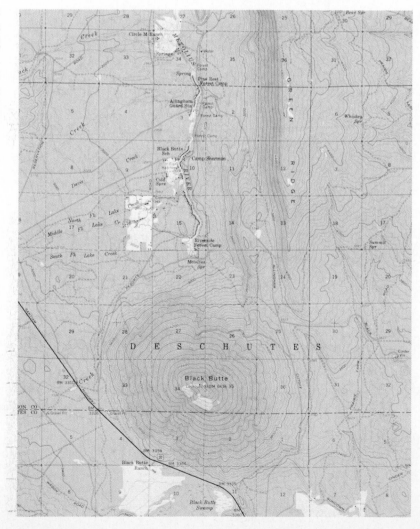

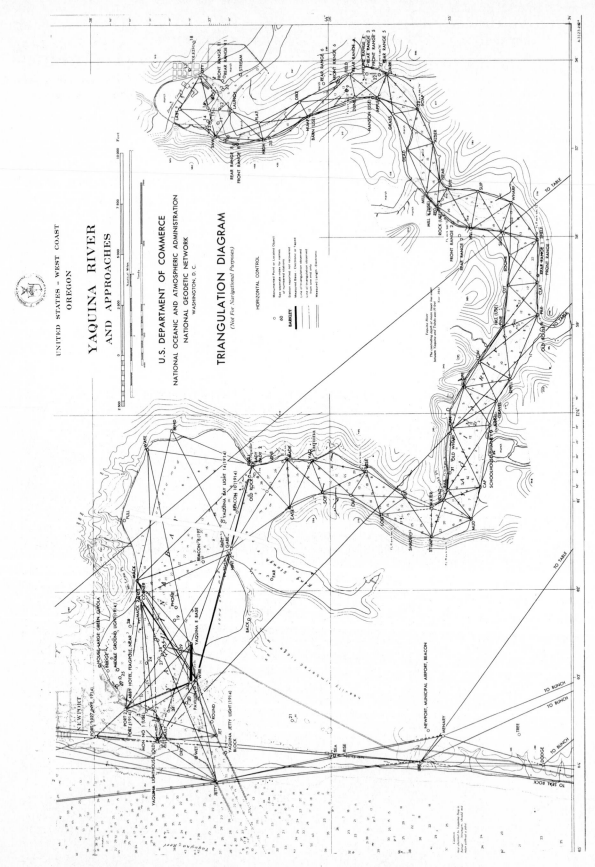

Figure 2-3 A geodetic survey defines major control points that can be used for smaller surveys. (Courtesy National Geodetic Survey)

called *triangulation*. A series of intersecting triangles is established as a net. Some sides of these triangles may be hundreds of miles long, stretching from one mountain peak to another. The control established by geodetic surveys is often used as references for other surveys. Figure 2-3 illustrates the size of a typical geodetic survey.

Photogrammetric Survey

Most topographic maps are now made using aerial photographs. Photographs taken at various altitudes constitute the "field notes" of this survey. Measurements are taken on the photos of known distances on the ground (often established by a land survey or open traverse) to check for accuracy, then maps are compiled using the information contained in the photograph. Many overlapping flights have to be flown before an accurate map can be created. An aerial photo used in photo mapping is shown in Figure 2-4.

Figure 2-4 Photogrammetric surveys produce aerial photos such as this. (Courtesy Spencer B. Gross Consulting Engineer)

Route Survey

We mentioned the term "open traverse" in the preceding paragraph and define it as a traverse that does not close on itself. An open traverse is the kind that is run when mapping linear features such as highways, pipelines, or power lines. These are termed route surveys. They can begin at a control point such as a bench mark and consist of straight lines and angles. These surveys do not close. A route survey is shown in Figure 2-5.

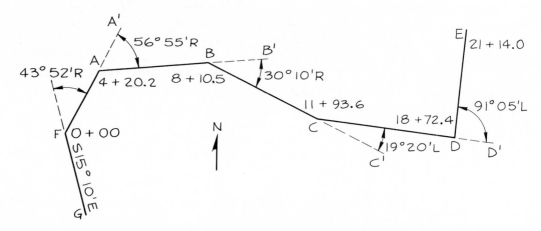

Figure 2-5   Route survey or open traverse.

Construction Survey

As the name implies, the construction survey is performed at construction sites. Building lines and elevations of excavations, fills, foundations, and floors are established by this localized type of survey (see Figure 2-6).

DISTANCE AND ELEVATION

The Chain

The roots of surveying were put down in ancient Egypt when ropes were used to measure taxable farmland on the Nile by men called "rope stretchers." From the rope came the chain and the steel tape. Whereas the steel tape, seen in Figure 2-7, normally stretches to 100 ft, the old *Gunter chain* measures 66 ft. Chains of 20-m lengths, termed *land chains* because of their use in land surveys, are also popular and convenient.

The use and measurements of a chain or tape is shown in Figure 2-8. Hubs or markers of some sort are placed at each point where a reading is to be made. When chain measurements must be made on a slope, the process is often referred to as *breaking chain,* and is shown in Figure 2-9. Most chaining is now done with the steel tape and plumb bobs.

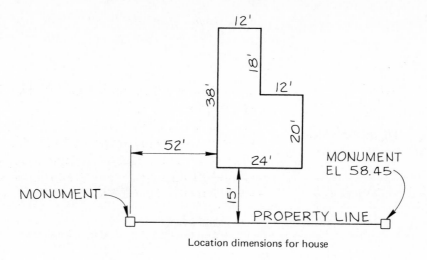

Location dimensions for house

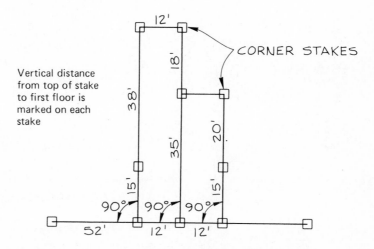

Figure 2-6 Construction survey showing locations of corners and staking out of house with angles and distances.

Figure 2-7 Steel tape used to measure distance. (Courtesy Lietz Company)

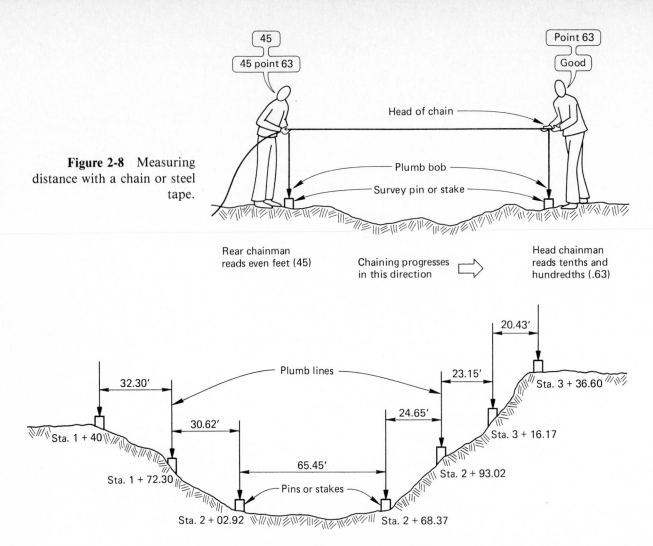

Figure 2-8 Measuring distance with a chain or steel tape.

45

45 point 63

Point 63

Good

Head of chain

Plumb bob

Survey pin or stake

Rear chainman reads even feet (45)

Chaining progresses in this direction ⟹

Head chainman reads tenths and hundredths (.63)

20.43'

32.30'

Plumb lines

23.15'

Sta. 3 + 36.60

30.62'

24.65'

Sta. 1 + 40

Sta. 3 + 16.17

65.45'

Sta. 1 + 72.30

Sta. 2 + 93.02

Pins or stakes

Sta. 2 + 02.92

Sta. 2 + 68.37

Figure 2-9 "Breaking chain" on sloping terrain. Chain must be horizontal for each measurement.

Electronics often take the legwork out of work, and surveying is no exception. Measuring distances with radar and lasers is a relatively simple matter once reflectors are set on distant tripods. Many companies employ instruments such as the distance meter shown in Figure 2-10. The accuracy is great, the time involved is less, and it is often a pleasure to work with precise, well-made equipment.

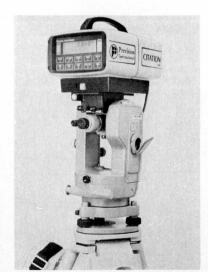

Figure 2-10 Theodolite with an electronic distance meter mounted on top. (Courtesy Wild Heerbrugg Instruments, Inc.)

Distance by Stadia

The Greeks used the word *stadium* (plural *stadia*) when referring to a unit of length. This unit was 600 Greek feet. That translates to 606 ft 9 in. in American feet. This unit of length was used when laying out distances in athletic contests. We now use the term to refer to a type of distance measuring employing a *rod* and an instrument with crosshairs.

The *Philadelphia rod* is 7 ft long, extends to 12 or 13 ft, and is graduated to hundreths of a foot, but can be read to thousandths. Distances are normally read only to hundreths. Elevations can be read to thousandths. Figure 2-11 illustrates the stadia method, which is based on optics. The space between two crosshairs in the instrument is read and multiplied by 100. This gives a fairly accurate distance; one that is sufficient for low-order surveys (those requiring a lower degree of precision).

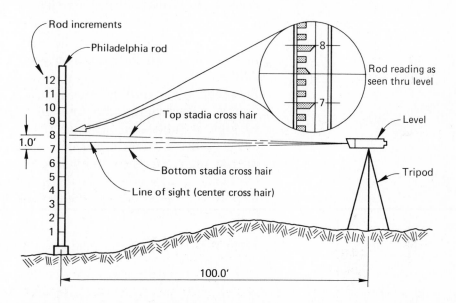

Figure 2-11　Measuring distance by "stadia theory."

Elevation by Level and Rod

The Philadelphia rod can also be used for measuring elevations when joined by an instrument called a level. *Leveling* is a process used to determine elevations of certain points and is done using a rod, level, and tripod. The rod is placed on the known elevation and the instrument is set up approximately halfway to the unknown point. Long distances may require several setups. The instrument is leveled and a reading is taken on the *backsight* (rod location). This is added to the elevation of the known point to give the *height of the instrument* (H.I.). The rod is then placed on the unknown elevation (*foresight*) and a reading is taken. This reading is subtracted from the height of the instrument to give us the unknown elevation (see Figure 2-12).

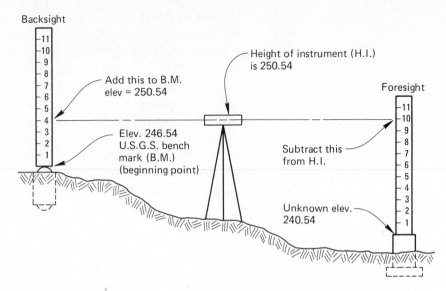

Figure 2-12 Finding an elevation with level and rod (leveling).

Measurements that must be made over long distances require *turning points,* which are nothing more than *temporary bench marks* (TBMs) and are often as temporary as a long screwdriver driven into the ground. The turning point is just a pivot for the rod, which is used as a backsight and a foresight. The instrument reading as a foresight is subtracted from the H.I. to find the level of the TBM and then the instrument and tripod are physically moved ahead of the rod and reset. The next reading is a backsight which is added to the TBM elevation to become the H.I. Now the rod can be moved to the next foresight position. Figure 2-13 illustrates leveling with several turning points.

Figure 2-13 Using turning points to find an unknown elevation.

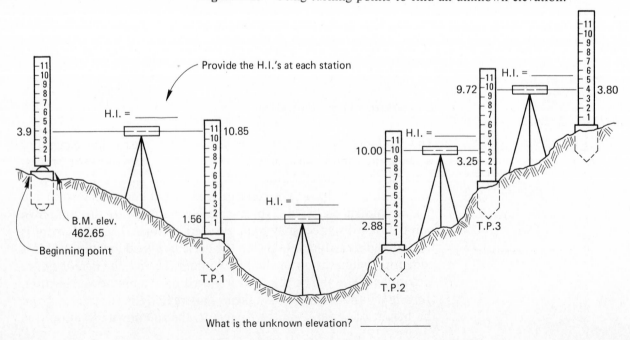

What is the unknown elevation? _____

TRAVERSING   A *traverse* is a series of continuous lines connecting points called *traverse stations* or *station points*. The lengths of the lines connecting the points are measured, as are the angles between the lines. Several traverse types are currently in use.

Open Traverse

The open traverse, as seen in Figure 2-14 and mentioned previously, consists of a series of lines that do not return to a POB and do not necessarily have to end on a control point. Most route surveys employ this type of traverse. The open traverse cannot be checked and is not suitable for work other than route surveys.

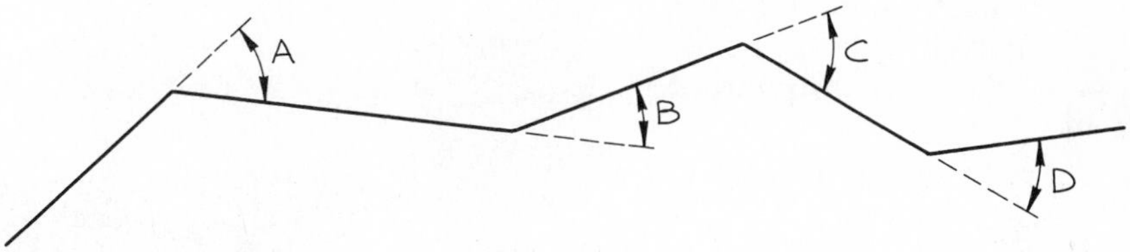

Figure 2-14

Closed Traverse

In the closed type of traverse, the lines close on the point of beginning, as in a *loop traverse,* or close on a different known control point, as in the *connecting traverse*. The closed traverse can be checked for accuracy and is thus used exclusively for land surveys and construction surveys. Figure 2-15 shows examples of loop and connecting traverses.

Figure 2-15   (a) Closed or loop traverse. (b) Connecting traverse.

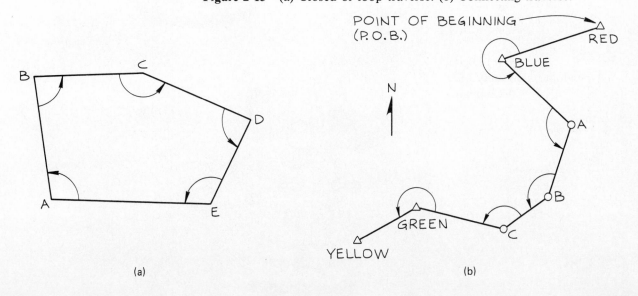

Compass Bearing Traverse

Bearings are angular measurements of 0 to 90 degrees taken from a north or south line and are oriented either east or west. The surveyors compass was originally used when laying out a traverse and the bearings were read directly from the compass. In a present-day compass bearing traverse, bearings are calculated using such instruments as the transit and theodolite (Figure 2-16). These bearings are more accurate, and can be used with great precision in calculations for mapping. The bearing of the backsight is known and the angle is then measured to the foresight. This angle is applied to the back bearing to determine the foresight bearing. This method is often used in connecting traverses, as shown in Figure 2-15b.

Figure 2-16 (a) Level transit. (Courtesy Berger Instruments)

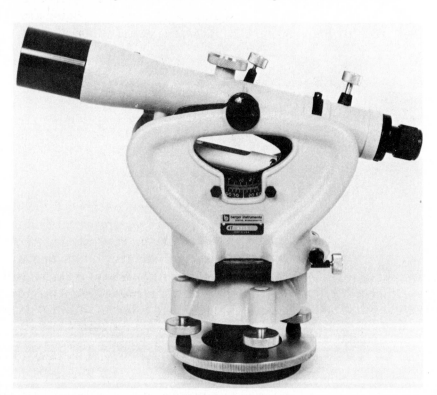

(b) Digital theodolite. (Courtesy Lietz Company)

Direct Angle Traverse

This is the principal method by which closed traverses are measured. The traverse proceeds either clockwise or counterclockwise around the plot and measures the interior angles. The closed traverse shown in Figure 2-15a is plotted by the direct angle method. Bearings can later be applied to the direct angles if required.

Deflection Angle Traverse

A deflection angle is one that veers to the right or left of a straight line (see Figure 2-17). This method of angle measurement is commonly used in route surveys. An angle to the right or left of the backsight is measured and is always 180° or less. The letters R or L must always be given with the angle.

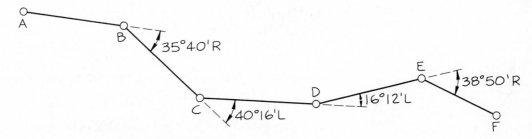

Figure 2-17   Deflection angle traverse.

Azimuth Traverse

Azimuth means horizontal direction. It is derived from the Arabian word "al-samt," which means "the way." In mapping, we use the term to refer to a direction that is measured from a north or south line. Unlike bearings, which measure only 90° quadrants, the azimuth is a measurement that encompasses the entire 360° of a circle. An azimuth traverse requires only one reference line and that is most often a north–south line. This reference line can be either true or magnetic. In Figure 2-18, the azimuths are measured clockwise from the north line.

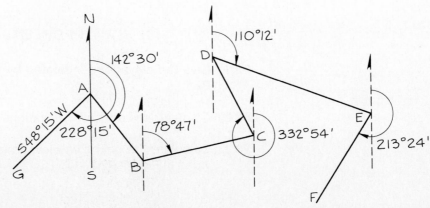

Figure 2-18   Azimuth traverse.

TEST

2.1. What is a topographic survey? *A DRAWING OF A PIECE OF LAND DEPICTING ITS ELEVATION AND FEATURES, BOTH NATURAL AND ARTIFICIAL*

2.2. Triangulation is a process used in this type of survey. *GEODETIC*

2.3. A route survey is often termed an open traverse. Why?
THEY DO NOT CLOSE

2.4. What are two things the Philadephia rod can be used for? *MEASURING ELEVATIONS WHEN JOINED BY A LEVEL*

2.5. What is "breaking chain"? *WHEN MEASUREMENTS ARE TO BE MADE ON A SLOPE*

2.6. Would a turning point be used to locate an elevation or distance? *YES*

2.7. What type of survey would be used to lay out a new highway? *CONSTRUCTION*

2.8. Can the type of survey in question 7 be checked easily? *NO*

2.9. What is a station point?

2.10. What type of instrument is used to measure bearings? *TRANSIT & THEODOLITE*

2.11. An angle to the right or left of the backsight is termed a *FORESIGHT*.

2.12. Compare and contrast bearings and azimuths.

PROBLEMS

P2.1. Give the elevations indicated by the rod readings in Figure P2-1.

P2.2. Calculate the unknown elevations in Figure P2-2.

P2.3. Calculate the unknown elevations given the field notes provided in Figure P2-3.

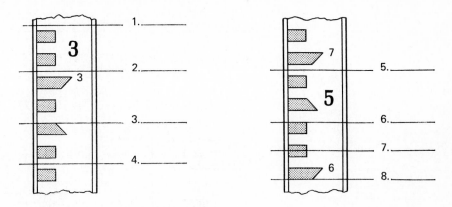

Figure P2-1

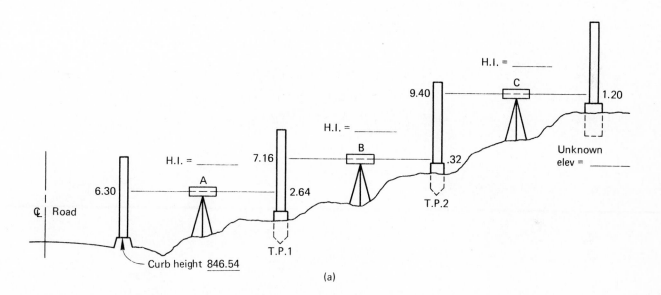

(a)

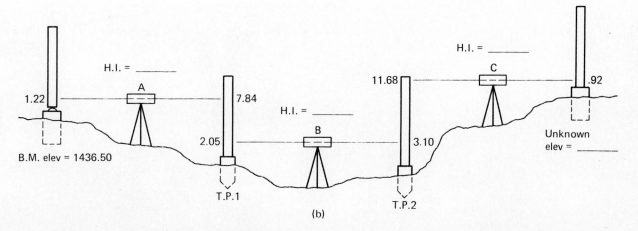

(b)

Figure P2-2

Station	Backsight (+)	H.I.	Foresight (−)	Elev
U.S.65.B.M.1				422.34
	6.21	428.55	3.46	
T.P.1	8.90		4.63	
T.P.2	7.82		2.45	
T.P.3 TOP OF CURB	5.60		3.86	
B.M.2 IRON PIN	12.04		7.70	
T.P.4	4.62		4.06	
T.P.5	7.31		3.64	

Figure P2-3

3

Location and Direction

Location and direction are two of the main purposes of mapmaking and use. This chapter explains the division of the earth into parts and how this system can be used to locate features. Also discussed are basic map geometry and construction of plats.

The topics covered include:
- Longitude
- Latitude
- Location on a map
- Direction
- Azimuth
- Bearing
- Map geometry
- Traverse

LOCATION The earth has been gridded by imaginary lines, called lines of *longitude* and *latitude*. These lines were established to aid location of features on a map.

If the earth were rectangular in shape, we could use a rectangular system that would be easy to measure and each square would be the same. With the earth "round" as it is, we have established points from which measurements can be accurately made. These points are the north and south poles, and the center of the earth. From these points, a grid system has been established using the degrees of a circle as reference. The grid lines are referred to as lines of longitude and latitude.

Longitude

Lines of longitude are imaginary lines that connect the north and south poles. These lines are also referred to as *meridians*. The imaginary line connecting the north and south poles, and passing through Greenwich, England, is called the *prime meridian*. The prime meridian represents 0° longitude. There are 180° west and 180° east of the prime meridian, forming the full circle of 360°. Meridians east of the prime meridian are referred to as *east longitude*. Meridians west of the prime meridian are called *west longitude*. The 180° meridian coincides roughly with the *International Date Line* (see Figure 3-1).

Length of a degree of longitude

A degree of longitude varies in length at different parallels of latitude, becoming shorter as the parallels approach the north and south poles. That is a little confusing, but think about it for a bit and refer back to Figure 3-1. Now you can see how degrees of longitude get closer together. Table 3–1 shows the length of a degree of longitude at certain latitudes.

Length of a Degree of Longitude

Latitude	Statute miles	Latitude	Statute miles
0°	69.171	50°	44.552
10°	69.128	60°	34.674
20°	65.025	70°	23.729
30°	59.956	80°	12.051
40°	53.063	90°	0.000

Figure 3-1 Measuring longitude.

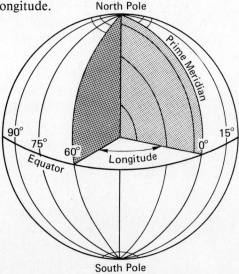

Latitude

Latitude is measured as an angular distance from the point at the center of the earth. Look at Figure 3-2 for an illustration. Notice that the latitude is the angle between the line of the equator and other side of the angle.

Points on the earth's surface that have the same latitude lie on an imaginary circle called a *parallel of latitude*. Lines of latitude are identified by degrees north or south of the equator. The equator is 0° latitude. The north pole is 90° *north latitude* and the south pole is 90° *south latitude*.

Figure 3-2 Measuring latitude.

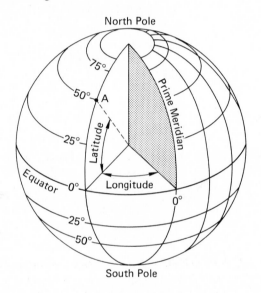

Length of a degree of latitude

Parallels of latitude are constructed approximately the same distance apart. Due to the bulging of the earth near the equator, each degree of latitude is not absolutely the same in distance. A degree of latitude varies from 69.4 statute miles near the poles to 68.7 statute miles at the equator. For all but the very detailed maps, it is satisfactory to refer to each degree of latitude as being 69 statute miles in length. (*Note: A statute mile* is established as an international standard and intended as a permanent rule.)

Location on Maps

On flat maps, meridians and parallels may appear as either curved or straight lines. Most maps are drawn so that north is at the top, south at the bottom, west at the left, and east at the right. Look for the north arrow for the exact orientation.

To determine directions or locations on any particular map, you must use the coordinates of parallels and meridians. So if you want to know which city is farthest north, Portland or Salem, the map in Figure 3-3 shows you that Portland is north of Salem. Which city is farthest east? Portland or Salem? The map tells the story. Salem is on the 123° meridian and Portland is east of that. The point that is actually being made is that any location on the earth can be identified by locating its intersecting lines of longitude or latitude.

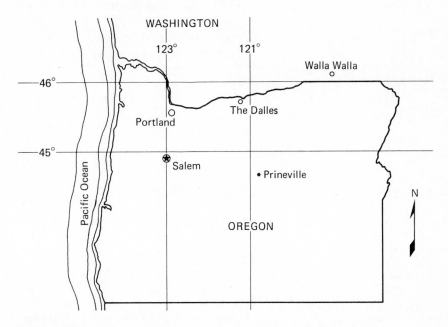

Figure 3-3 Using coordinates of parallels and meridians to find locations.

DIRECTION A *direction*, in surveying, refers to the angular relationship of one line to another. When a number of lines radiate from a point, the direction of these lines is expressed with reference to one of the lines that is designated as having zero direction. In most cases a north-south, or east-west line carries a zero designation.

Units of Angular Measure

Units of angular measure are degrees. Degrees are identified with the symbol °, as in 30°. There are 360° in a complete circle, hence a quarter circle has 90°, and so on. Each degree is made up of 60 minutes. The minutes symbol is ′. Minutes are divided into 60 seconds, which are identified as ″. So a complete degree, minute, and second designation may read like this: 50° 30′ 45″. It is important when you are doing calculations with degrees and parts of degrees that you keep this information in mind. Consider these examples:

$$48°40′25″$$
$$+\ 25°38′40″$$
$$\overline{73°78′65″}$$

$$\text{borrow 1°}\quad 74°60′$$
$$75°32′10″$$
$$-\ 34°45′\ 8″$$
$$\overline{40°47′\ 2″}$$

reduced to 74° 19′ 5″

Surveyor's Compass

The surveyor's compass is used in mapping to calculate the direction of a line. The reading achieved is usually a bearing angle or an included angle. Figure 3-4 shows the differences between two commonly used compasses: the mariner's compass and the surveyor's compass.

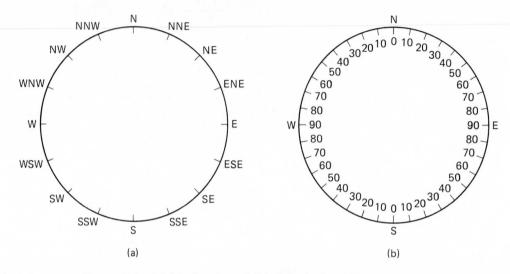

Figure 3-4   (a) Mariner's and (b) surveyor's compass.

Azimuth

An azimuth is a direction, measured as a horizontal angle from a zero line, generally north–south, in a clockwise direction (see Figure 3-5).

Figure 3-5   Azimuth.

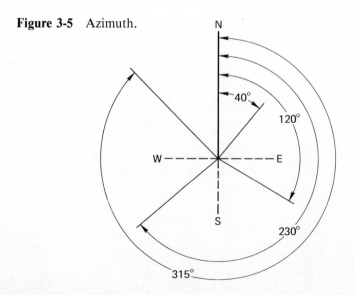

True Azimuth

A true azimuth is a horizontal angle measured using true north as the reference line.

Grid Azimuth

A grid azimuth is established for a rectangular survey system so that the north–south grids of the survey are used as the reference, or zero line. You can see this used when you work with the township section system.

Magnetic Azimuth

Magnetic azimuths are measured with magnetic north as the zero line. Actually, the magnetic compass is used only as a check on more accurate methods and as a method to obtain approximate values for angles. The magnetic azimuths may differ from the true azimuth by several degrees, depending on the local magnetic attractions.

Bearings

The bearing of a line is its direction with respect to one of the quadrants of the compass. Bearings are measured clockwise or counterclockwise, depending on the quadrant, and starting from north or south (see Figure 3-6).

A bearing is named by identifying the meridian, north or south; the angle; and the direction from the quadrant, east or west. Therefore, a line in the northeast quadrant with an angle of 60° with the north meridian will have a bearing of N60°E. Consider a line in the northwest quadrant that is 40° from north; the bearing will be N40°W. Look at the other examples in Figure 3-7.

Figure 3-6 Bearing.

Figure 3-7 Examples of bearing.

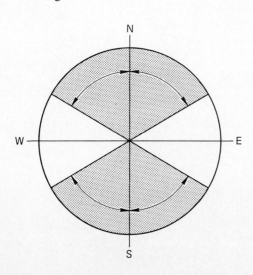

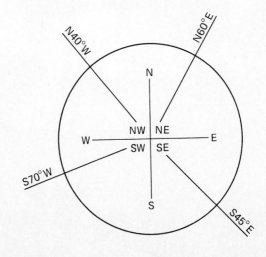

Magnetic Declination

The meridian indicated by the needle of a magnetic compass will seldom coincide with the true meridian. So the horizontal angle between the magnetic meridian and the true meridian at any point is called the magnetic declination. The magnetic declination is either east or west, depending on the direction the arrow of the compass points from true north. The magnetic declination of a map must be updated periodically because of continuous changes in its value. Here is how you determine a true azimuth when given the magnetic declination:

1. East magnetic declination + magnetic azimuth = true azimuth.
2. Magnetic azimuth − magnetic declination = true azimuth.

Figure 3-8 presents an example of magnetic declination. Figure 3-8 also shows an example of grid north and magnetic north in relationship to true north as taken from the legend of an actual map.

When a drafting technician draws bearing lines from the surveyor's notes, they are either true bearings or magnetic bearings. If they are true bearings, the surveyor notes in his field book: "Bearings are referred to the true (north–south) meridian." The bearings are corrected by applying the magnetic declination for the correct time (year and day). The drafting technician plots the corrected bearings as given in the field notes. The completed map should indicate if the bearings are true bearings.

Most maps carry the magnetic declination for a specific year, as well as the amount of annual change in degrees. The change is indicated as easterly or westerly, and records whether the compass is

Figure 3-8 Samples of magnetic declination.

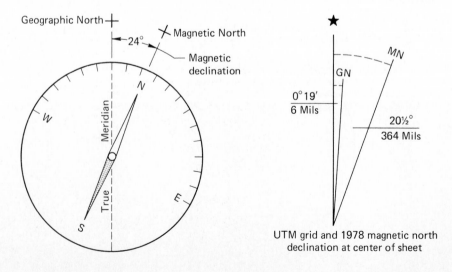

UTM grid and 1978 magnetic north
declination at center of sheet

pointing east of true north (easterly), or west of true north (westerly). After calculating the annual change, the amount must be either added to or subtracted from the angle of declination.

Local Attraction

A local attraction is any local influence that causes the magnetic needle to deflect away from the magnetic meridian.

Figure 3-9 shows various lines labeled with their relative bearings and azimuths.

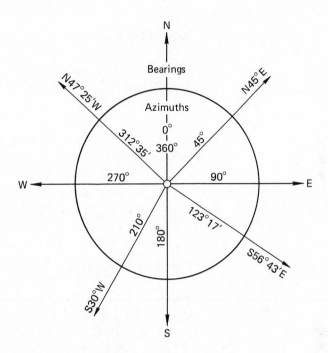

Figure 3-9 Examples of bearings and azimuths.

MAP GEOMETRY The information that you have learned concerning longitude, latitude, azimuth, and bearing can be put to use in the construction of plats or plots of land. Plots of property are drawn with border lines showing the starting point for location, bearing for direction, and dimensions for size. The following discussion shows how these three components are put together to form the boundaries for a plot.

Polygon

In plotting a traverse you remember that a closed traverse is a polygon. A polygon will close if all included angles equal 360°, as in Figure 3-10.

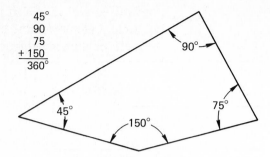

```
 45°
 90
 75
+150
 360°
```

Figure 3-10 All angles of a polygon will equal 360° when added together.

Intersecting Lines

When two lines intersect, the opposite angles are equal. In Figure 3-11, angle $X = X$ and $Y = Y$.

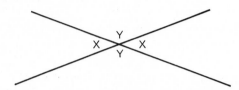

Figure 3-11 Opposite angles of intersecting lines are equal.

Plotting a Traverse

These concepts become important when you attempt to plot a traverse without all bearings given.

You may find a situation similar to this example. Given angles *A, B, C,* and *D,* plot the traverse and determine the bearings; be sure you understand what a bearing is. Make your calculations in a clockwise direction beginning at the point of beginning (POB). Look at Figure 3-12.

1. Let us begin by determining the bearing of one line at a time, first line *AB.* The bearing of line *AB* is due north. You can reason this because line *AB* is on the section line that is parallel to the north–south line.

2. Next we determine the bearing of line *BC* (see Figure 3-13). The bearing is in the northeast quadrant, so it would be N69°22′E.

3. We now calculate the bearing of *CD* (see Figure 3-14). We first must consider everything that we know.

   a. Angle *BCD* = 111°49′.

   b. Angle *X* = 69°22′, the bearing of line *BC* that we just calculated.

Line	Dist.	Angle	Included degrees
AB	170.1'	DAB	69°10'
BC	131.2'	ABC	110°38'
CD	173.2'	BCD	111°49'
DA	255.0'	CDA	68°23'

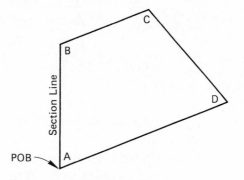

Figure 3-12 Example of a typical traverse. Included angles and distances given.

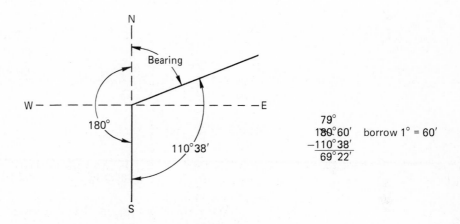

79°
1̶8̶0̶°̶60'
−110°38'
‾‾‾‾‾‾‾
69°22'   borrow 1° = 60'

Figure 3-13 Calculating bearings: line *BC*.

Figure 3-14 Calculating bearings: line *CD*.

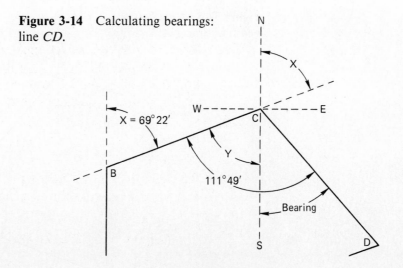

c. Angle X = angle X, because if two parallel lines are cut by another line, the exterior–interior angles on the same side of the line are equal.

d. Angle X = angle Y, because when one straight line intersects another, the opposite angles are equal.

e. Finally, we have concluded that $Y = 69°22'$, so the bearing will be

$$
\begin{array}{r}
111°49' \\
-\ \ 69°22' \\
\hline
42°27'
\end{array}
$$

The bearing is in the southeast quadrant, so it would be S42°27'E.

4. Finally, we determine the bearing of DA, and we will be back to the POB (see Figure 3-15).

We know from the previous part that angle $Z = 42°27'$. We also know from our elementary geometry that $Z = Z_1 = Z_2$. So

$$
\begin{array}{r}
68°23' \\
+\ \ 42°27'\ (Z_2) \\
\hline
110°50'
\end{array}
$$

Now,

$$
\begin{array}{r}
179°60' \\
-\ \ 110°50' \\
\hline
\text{Bearing} \qquad 69°10'
\end{array}
$$

With this bearing in the southwest quadrant, the reading is S69°10'W. Your complete plat layout should appear as shown in Figure 3-16.

Figure 3-15 Calculating bearings: line DA.

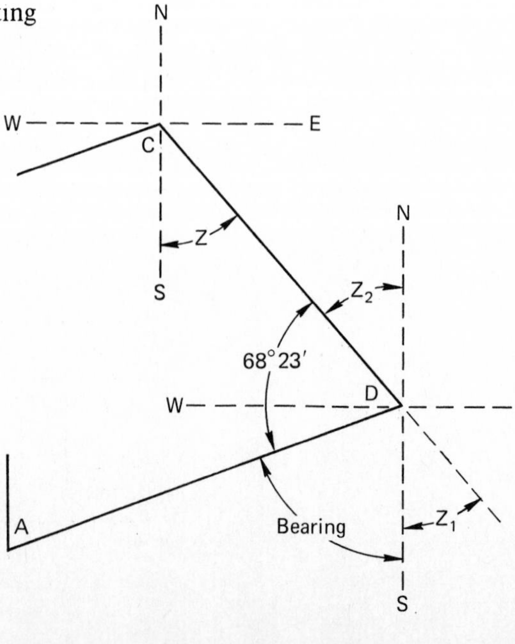

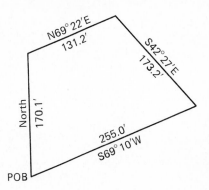

Figure 3-16 Bearings and distances shown on a plat.

Traverse

Plotting a traverse is a method of checking the accuracy of a survey. This consists of a series of lines and angles from the survey. Starting at a given point, these angles and lines are plotted to form a closed polygon. When you, in fact, do close the polygon, you know that the traverse is accurate, and you have what is called a *closed traverse*. Figure 3-17 is a surveyor's rough sketch of a surveyed plat. The sketch includes all line lengths and bearings. You may be given azimuths or interior angles; can you convert to bearings? An open traverse may occur when the ends do not intend to close.

Another way to present the information is in a plotting table such as the one shown in Table 3-2. If you are given a plotting table, the information may not be on the sketch.

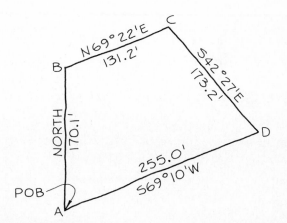

Figure 3-17 Rough sketch of a plat.

Table 3-2 Plotting Table

Line	Bearing	Distance	Angle
AB	North	170.1′	69°10′
BC	N69°22′E	131.2′	110°38′
CD	S42°27′E	173.2′	111°49′
DA	S69°10′W	255.0′	68°23′

Error of Closure

When a surveyor expects to get a closed traverse and does not quite make it, he or she can cure the problem using the error of closure method. As a drafting technician, you may also have an error due to the small inaccuracies involved in layout. These errors may result in a plot that does not quite close; one such plot is shown in Figure 3–18.

Here is what you do:

1. Lay out the perimeter of the plot along a straight line and label the corners, as in Figure 3-19.

2. Extend the error of closure above point A'.

3. Connect a line from A'' to A.

4. Now connect perpendicular lines from points B, C, and D to line A–A''.

5. You have the distance that you can allow at each point to compensate for the problem. Notice how this is done in Figure 3-20.

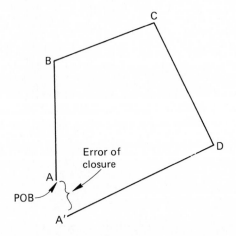

Figure 3-18 Error of closure.

Figure 3-19 Correcting the error of closure.

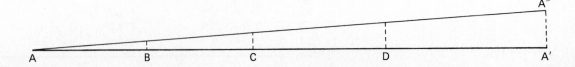

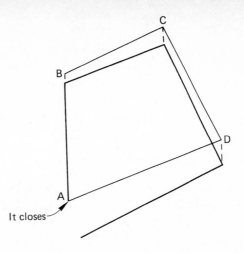

Figure 3-20 Closed traverse.

TEST

Part I

Multiple choice: Circle the response that best describes each statement.

3.1. The line that represents zero degrees longitude is also called:
 a. The International Date Line
 b. The prime meridian
 c. The equator
 d. Greenwich, England

3.2. Lines of latitude:
 a. Are measured as an angular distance from the point at the center of the earth
 b. Connect the north and south poles
 c. Are called parallels
 d. Are approximately the same distance apart

3.3. 45°26′14″ + 15°10′52″ =
 a. 60°36′6″
 b. 60°37′66″
 c. 60°37′6″
 d. 61°13′6″

3.4. A direction measured clockwise from a given zero is called:
 a. A bearing
 b. An azimuth
 c. A magnetic declination
 d. A local attraction

3.5. A direction measured clockwise or counterclockwise with respect to quadrants of a compass is called:
 a. A grid azimuth
 b. An azimuth
 c. A bearing
 d. A magnetic declination

Part II

3.1.   Give the azimuths of the following lines. North is zero.

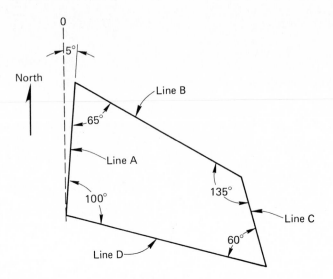

        Line *A* _____
        Line *B* _____
        Line *C* _____
        Line *D* _____

3.2.   Give the bearings of the lines featured in the question above.

        Line *A* _____
        Line *B* _____
        Line *C* _____
        Line *D* _____

Part III

True or false: Circle the "T" if the statement is true or the "F" if the statement is false. Reword all false statements so that the meaning is true.

3.1. ⓣ F   Magnetic azimuths may differ from true azimuth by several degrees.

3.2. ⓣ F   The horizontal angle between the magnetic meridian and the true meridian is called the magnetic declination.

3.3. T Ⓕ   Magnetic azimuth + magnetic declination = true azimuth.

3.4. T Ⓕ   The magnetic declination does not change.

3.5.   T   (F)   The included angles of a polygon equal 360°.

3.6.   T   (F)   When drawing a plat based on angular information, you must convert bearings to azimuths or interior angles.

Part IV

Given the following groups of information, determine the distances on the earth's surface. Use your knowledge of lengths of a degree of longitude and latitude. Make sketches of the earth showing the points given to help you establish location.

3.1.   The distance between a point at 45° north latitude, 20° west longitude and a point 30° north latitude, 20° west longitude.

3.2.   The distance between a point at 16° north latitude, 60° east longitude and a point at 28° south latitude, 60° east longitude.

3.3.   The distance between a point at 50° north latitude, 80° west longitude and a point at 50° north latitude, 20° west longitude.

3.4.   The distance between a point at 10° north latitude, 80° west longitude and a point at 10° north latitude, 20° west longitude.

3.5.   The distance between a point at 72° north latitude, 28° west longitude to a point at 18° north latitude, 28° west longitude to a point at 18° north latitude, 42° west longitude.

PROBLEM

P3.1.   Tape Figure P3-1 to your drawing board. With your drafting machine protractor or hand-held protractor, determine the azimuth and bearing for each line. Place your answer in the table provided. You will be evaluated on accuracy to ± 15′ and on the neatness of your lettering in the table.

Line	Azimuth	Bearings
A		
B		
C		
D		
E		
F		
G		
H		
I		

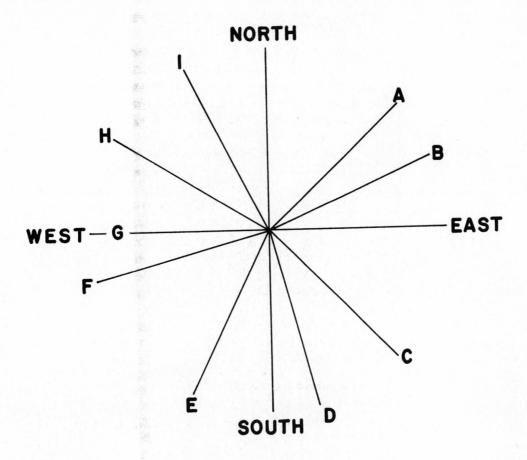

Figure P3-1

4

Mapping Scales

This chapter explains the purpose and types of map scales. Illustrations are provided to show how scales affect the map that is being represented.

The topics covered include:

- Numerical scale
- Graphic scale
- Verbal scale
- Scale conversion
- Use of the engineer's scale

Map scales vary as to purpose, size, and desired detail. For example, if a map of the State of New York were to be painted to scale on a large, blank wall of a building, the scale would be different than if the New York map were drawn to scale on an 11- by 17-in. sheet of paper. Also, a map scale will vary according to the area of the earth's surface covered. A map of your school drawn on this page will have a different scale than if it were a map of the United States.

A simple definition of a map is a representation of a portion of the earth's surface reduced to a small size. The map's scale aids you in estimating distances, and gives you an idea of what to expect in the way of detail. The map scale shows the relationship between the measurements of the features shown on the map compared to the same features on the earth's surface.

There are three methods of expressing scale:

1. Numerical scale or representative fraction
2. Graphic scale
3. Verbal scale

NUMERICAL SCALE The numerical scale gives the proportion between the length of a line on a map and the corresponding length on the earth's surface. This proportion is known as the *representative fraction* (RF). The first number is a single unit of measure equal to the distance on the map = 1. The second number is the same distance on the earth using the same units of measure. It can be written in two ways: 1/150,000 or 1:150,000.

RF (representative fraction) = distance on map/distance on earth

It is very important to note that this map scale must always refer to map and ground distances in the same unit of measure. The unit of measure most commonly used in the representative fraction is the inch. For instance, a scale of 1 in. on your map representing 1 mile on the earth's surface would be expressed as follows: RF = 1/63,360 or 1:63,360. (One mile has 63,360 in.)

Remember, the numerator is always 1, and represents map. The denominator is always greater, because it represents ground.

To calculate the distance between two points on a map, we must multiply the measured distance in inches by the denominator of the RF. Let us presume that we have measured 2¼ in. between *A* and *B* on a map drawn to a scale of 1/100,000. The distance between points *A* and *B* would be: distance = 2¼ × 100,000 = 225,000 in. You can convert this to feet or miles if you wish. If the map were calibrated in metrics, 1:100,000 would mean 1 cm = 100,000 cm = 1 km.

Small-Scale and Special Maps

Information compiled in a pamphlet titled *Tools for Planning* by the U.S. Department of Interior, Geological Survey, indicates that several series of small-scale and special maps are published to meet modern requirements. These include:

- *1:250,000 scale* (1 in. on the map represents about 4 miles on the ground)—for regional planning and as topographic bases for other types of maps.
- *State maps*, at 1:500,000 scale (1 in. represents 8 miles)—for use as wall maps and for statewide planning.

- *Shaded-relief maps*, at various scales—for park management and development and as tourist guides.
- *Antarctic topographic maps*, at various scales—for scientific research.
- *International map of the world* (IMW), at 1:1,000,000 scale—for broad geographic studies.

Figure 4-1 provides a graphic example of the previous discussion.

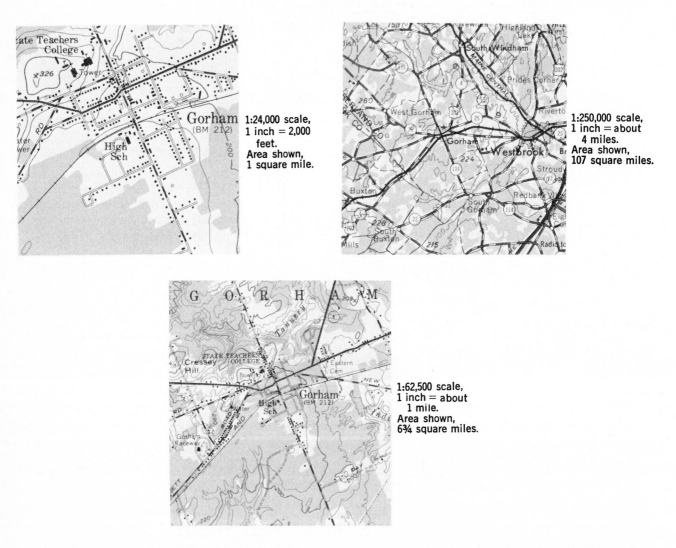

1:24,000 scale, 1 inch = 2,000 feet. Area shown, 1 square mile.

1:250,000 scale, 1 inch = about 4 miles. Area shown, 107 square miles.

1:62,500 scale, 1 inch = about 1 mile. Area shown, 6¾ square miles.

Figure 4-1 Reproduced from "Topographic Maps" by permission of the U.S. Geological Survey.

GRAPHIC SCALE This kind of map scale is like a small ruler in the legend or margin of the map. The divisions on the graphic scale represent increments of measure easily applied to the map.

Ordinarily, graphic scales begin at zero, but many have an extension to the left of the zero. This enables you to determine distances less than the major unit of the scale. A graphic scale uses relatively even units of measure, such as 500 or 1000. Figure 4-2 illustrates three graphic scales, each calibrated differently.

Suppose that you wish to measure the distance down the road between points *A* and *B*. Just apply a straight-edged strip of plain paper on the map along the distance between the two points and make marks on the paper strip where the edge touches *A* and *B*. Then move the strip of paper to the graphic scale and measure. Point *A* is found to be 1.5 miles down the road from point *B*. Figure 4-3 illustrates how the piece of paper is used to mark the distance on the map and then determine the distance measured.

Figure 4-2 Sample graphic scales.

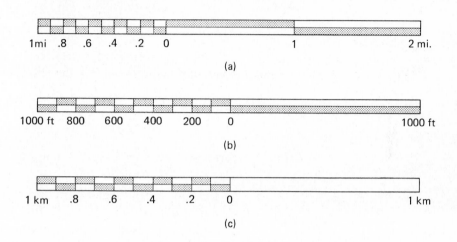

Figure 4-3 Measuring with a graphic scale.

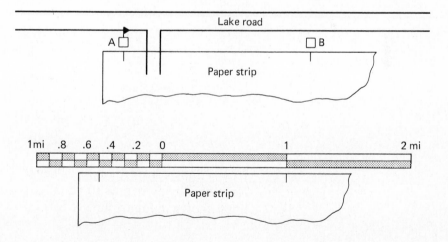

VERBAL SCALE   Verbal scale is usually expressed in the number of inches to the mile. Usually, the mileage is rounded. For example, 1 in. on an RF scale of 1:1,000,000 equals 15.78 miles. This distance expressed in verbal scale would round off the 15.78 to 16. So the verbal scale would be 1 in. = 16 miles. This is a close approximation for discussion purposes. The verbal scale is not meant to be accurate, only a rough estimate. Use the RF scale for accurate measurements.

SCALE CONVERSION

In map drafting or map reading it is often necessary to convert from one scale to another. You may have to convert an RF scale to a graphic scale for display on the map.

Convert a Representative Fraction to a Graphic Scale

Example: Given a representative scale of 1:300,000, find the graphic equivalent. Remember that the 1 equals 1 in. on the map and the 300,000 equals 300,000 in. on the earth's surface. We must first determine the number of miles that are represented by the 300,000 in. There are 63,360 in. in a mile, so

$$\frac{300,000 \text{ in.}}{63,360 \text{ in.}} = 4.73 \text{ miles}$$

This scale is not practical because it is difficult to imagine odd measurements such as 4.73 miles (Figure 4-4).

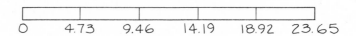

Figure 4-4 This graphic scale is not practical.

Instead, graphic scales are drawn using units like 0–1–2–3–4 or 0–2–4–6 or 0–5–10–15–20. To do this, it is necessary to find out how many inches are used to show any even number larger than the 4.73 miles previously determined. Let us choose 10 miles. This would be set up in a ratio:

$$\frac{4.73 \text{ miles}}{1 \text{ in.}} = \frac{10 \text{ miles}}{X \text{ (unknown no. of inches)}}$$

Use cross-multiplication to establish this algebraic formula:

$$4.73X = 10$$
$$X = 2.11 \text{ in.}$$

In 2.11 in. we have 10 miles. Then it is easy to lay off a line 2.11 in. long and divide it into 10 equal parts using the method of parallel lines, shown in Figure 4-5.

Figure 4-5 Correct graphic scale with one segment divided into 10 equal spaces.

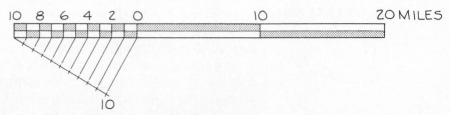

Convert a Graphic Scale to a Representative Fraction

Example: 1 in. on the graphic scale equals 5 miles. (1 mile = 63,360 in.)

$$5 \times 63{,}360 = 316{,}800$$
$$RF = 1{:}316{,}800$$

Determine the representative fraction for a map with no scale.

The representative fraction can be determined by measuring on a straight meridian line the distance for the map's 1° of latitude. The actual average distance in 1° of latitude is 69 miles. So if the number of inches between 1° of latitude equals 2.3, the following formula will work to solve the RF:

$$RF = \frac{\text{miles in 1° of latitude} \times 63{,}360 \text{ in.}}{\text{inches in 1° of latitude on the map}}$$

$$= \frac{69 \times 63{,}360}{2.3}$$

$$= 1{:}1{,}900{,}800$$

ENGINEER'S SCALE The maps that you draw will require a scale so that the reader will be able to interpret distances accurately. Different methods of expressing map scales have been discussed. Now, you need a tool to use that will allow you to draw a map at a scale that you select. Commonly, maps are drawn using a scale that is made up in multiples of 10. For example, 1 in. = 10 ft, 100 ft, 1000 ft, 10,000 ft, and so on. The tool that is often used by the drafting technician is an engineer's scale. Figure 4-6 shows an example of a typical triangular engineer's scale.

There are six engineer's scales found on the triangular engineer's scale. Each scale is a multiple of 10, and may be used to calibrate a drawing in any units, such as feet, meters, miles, or tenths of any typical unit.

Figure 4-6 (a) A typical triangular engineer's scale. (b) Flat engineer's scale.

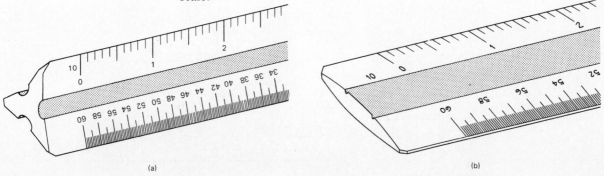

This list shows the six possible scales:

10 scale: 1 in. = 1.0 ft; 1 in. = 10.0 ft; 1 in. = 100.0 ft; etc.
20 scale: 1 in. = 2.0 ft; 1 in. = 20.0 ft; 1 in. = 200.0 ft; etc.
30 scale: 1 in. = 3.0 ft; 1 in. = 30.0 ft; 1 in. = 300.0 ft; etc.
40 scale: 1 in. = 4.0 ft; 1 in. = 40.0 ft; 1 in. = 400.0 ft; etc.
50 scale: 1 in. = 5.0 ft; 1 in. = 50.0 ft; 1 in. = 500.0 ft; etc.
60 scale: 1 in. = 6.0 ft; 1 in. = 60.0 ft; 1 in. = 600.0 ft; etc.

Along the margin of the engineer's scale, you will see a 10 on one edge. This 10 represents the 10 scale. Another edge will have a 20 in the margin representing the 20 scale. This same situation holds true for the 30, 40, 50, and 60 scales. Figure 4-7 is a close-up of the margin on an engineer's scale.

Measurements are easy to multiply and divide since they are given in decimals rather than in feet and inches. Figure 4-8 provides some excellent illustrations of how the engineer's scale may be used to measure distances.

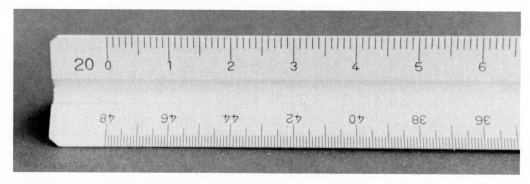

Figure 4-7   Margin of triangular engineer's scale.

Figure 4-8   Three examples of margins and scales on an engineer's scale.

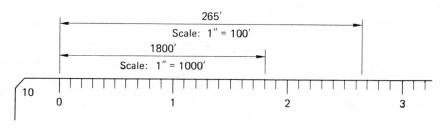

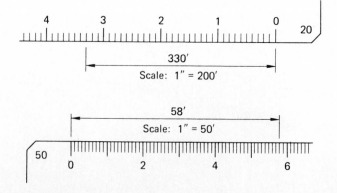

TEST

Part I

Fill in the blanks below with the responses necessary to complete each statement. Use your best freehand lettering.

4.1.   A __MAP__ is a representation of a portion of the earth's surface reduced to a small size.

4.2.   The proportion between the length of a line on a map and the corresponding length on the earth's surface is called the __REPRESENTATIVE FRACTION__.

4.3.   A scale used to approximate distances on a map, that usually looks like a small ruler found near the legend or title block, is called the __GRAPHIC__ __SCALE__.

4.4.   Representative fraction = RF = $\dfrac{\text{DISTANCE ON MAP}}{\text{DISTANCE ON EARTH}}$.

4.5.   A scale that is often used for rough estimates of distances on a map is called the __VERBAL__ __SCALE__.

Part II

Make the following scale conversions. Be neat and accurate. Show your calculations in the space provided.

4.1.   Convert RF = 1:100,000 to a graphic scale.

4.2.   Convert RF = 1:2000 to a graphic scale.

4.3.   Convert a verbal scale of 1 in. = 1 mile to a representative fraction.

4.4.   Convert a verbal scale of 1 in. = 100 miles to RF.

4.5.   If the number of inches between 1° of latitude on a map is measured to be 4.75 in., what is the RF of the map?

Part III

Use your civil engineer's scale to measure each line shown below. The scale calibration is identified to the left of each line and place your answer to the right.

	Scale		Answer
4.1.	1 in. = 10 ft	_____	19'6"
4.2.	1 in. = 20 ft	_____	39'
4.3.	1 in. = 30 ft	_____	58'6"
4.4.	1 in. = 40 ft	_____	78'
4.5.	1 in. = 50 ft	_____	97'6"
4.6.	1 in. = 60 ft	_____	117'
4.7.	1 in. = 2 ft	_____	_____
4.8.	1 in. = 100 ft	_____	195'

PROBLEM

P4.1. Convert the engineer's sketch in Figure P4-1 to a formal drawing. Remember, the sketch is not accurate, so you will have to lay out all bearings and distances. Do not trace the sketch. Consider the following.

1.   Use B-size vellum 11 × 17 in.
2.   Use pencil.

3. Include:
    a. Title
    b. Scale 1 in. = 20 ft
    c. Graphic scale
    d. North arrow
    e. Your name and date
4. Make a print for instructor evaluation unless otherwise specified.

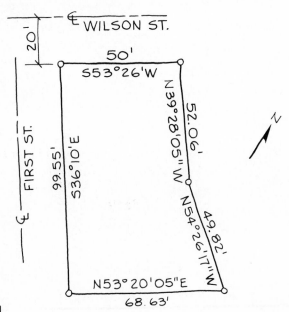

Figure P4-1

5

Mapping Symbols

This chapter discusses and shows examples of some of the symbols that are commonly used in civil drafting. These are only a few of the actual number of symbols that are available for your use. The purpose of the map largely dictates the symbols to be used. Special symbols sometimes have to be designed for a specific purpose. The symbols shown may change slightly. Symbols may include colors or may be in black and white.

The topics covered include:

- Types of symbols
- Symbol colors, when used
- Special-effect symbols

Standard symbols are used in civil drafting as they are in any of the other fields of drafting. When you look at a road map, for example, you generally take it for granted that symbols on the map will help you get where you want to go. Without these symbols, the map would be useless. Each type of map may have certain symbols that are special to the intent of the map. Specific symbols are also used to help keep information on a map to a minimum so that the map remains as uncluttered as possible. Symbols are as condensed and to the point as possible to describe the objective.

TYPES OF SYMBOLS   The various symbols used in map drawings can be grouped under four types: culture, relief, water, and vegetation.

Culture

Culture symbols represent works of people. When maps are printed in color, these symbols are usually done in black. Lettering on maps that identify cultural representation is usually done in vertical caps. Variations will be found to standards that exist. The topographic map symbols shown in Figure 5-1 show several different cultural symbols.

Figure 5–1 Standard topographic map symbols. Reproduced by permission of the U.S. Geological Survey.

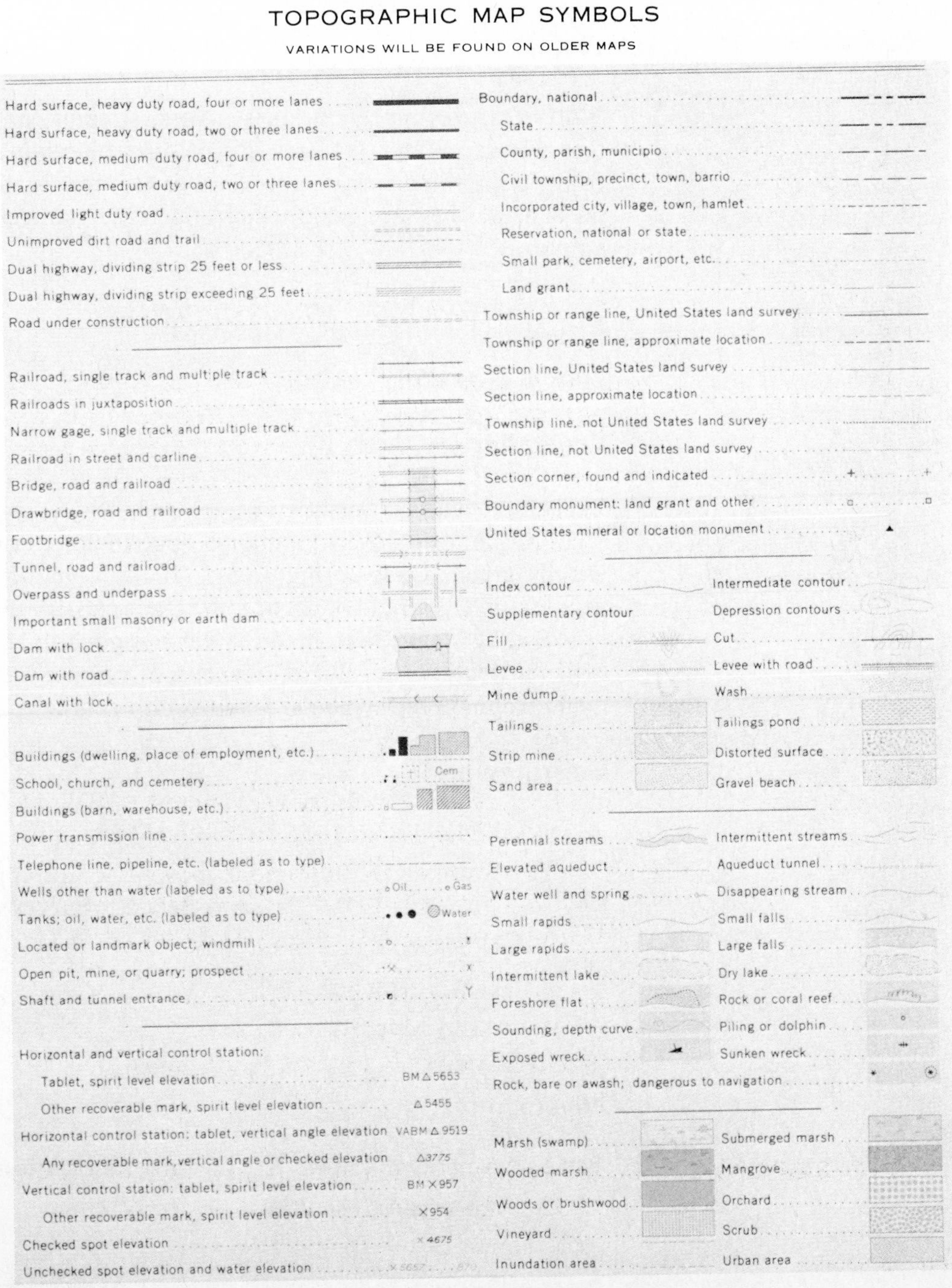

TOPOGRAPHIC MAP SYMBOLS
VARIATIONS WILL BE FOUND ON OLDER MAPS

Hard surface, heavy duty road, four or more lanes

Hard surface, heavy duty road, two or three lanes

Hard surface, medium duty road, four or more lanes

Hard surface, medium duty road, two or three lanes

Improved light duty road

Unimproved dirt road and trail

Dual highway, dividing strip 25 feet or less

Dual highway, dividing strip exceeding 25 feet

Road under construction

Railroad, single track and multiple track

Railroads in juxtaposition

Narrow gage, single track and multiple track

Railroad in street and carline

Bridge, road and railroad

Drawbridge, road and railroad

Footbridge

Tunnel, road and railroad

Overpass and underpass

Important small masonry or earth dam

Dam with lock

Dam with road

Canal with lock

Buildings (dwelling, place of employment, etc.)

School, church, and cemetery

Buildings (barn, warehouse, etc.)

Power transmission line

Telephone line, pipeline, etc. (labeled as to type)

Wells other than water (labeled as to type)   o Oil   o Gas

Tanks; oil, water, etc. (labeled as to type)   • • ● ⊙ Water

Located or landmark object; windmill

Open pit, mine, or quarry; prospect

Shaft and tunnel entrance

Horizontal and vertical control station:

Tablet, spirit level elevation   BM △ 5653

Other recoverable mark, spirit level elevation   △ 5455

Horizontal control station: tablet, vertical angle elevation VABM △ 9519

Any recoverable mark, vertical angle or checked elevation   △3775

Vertical control station: tablet, spirit level elevation   BM × 957

Other recoverable mark, spirit level elevation   × 954

Checked spot elevation   × 4675

Unchecked spot elevation and water elevation   × 5657

Boundary, national

State

County, parish, municipio

Civil township, precinct, town, barrio

Incorporated city, village, town, hamlet

Reservation, national or state

Small park, cemetery, airport, etc.

Land grant

Township or range line, United States land survey

Township or range line, approximate location

Section line, United States land survey

Section line, approximate location

Township line, not United States land survey

Section line, not United States land survey

Section corner, found and indicated   +   +

Boundary monument: land grant and other   □   □

United States mineral or location monument   ▲

Index contour   Intermediate contour

Supplementary contour   Depression contours

Fill   Cut

Levee   Levee with road

Mine dump   Wash

Tailings   Tailings pond

Strip mine   Distorted surface

Sand area   Gravel beach

Perennial streams   Intermittent streams

Elevated aqueduct   Aqueduct tunnel

Water well and spring   Disappearing stream

Small rapids   Small falls

Large rapids   Large falls

Intermittent lake   Dry lake

Foreshore flat   Rock or coral reef

Sounding, depth curve   Piling or dolphin

Exposed wreck   Sunken wreck

Rock, bare or awash; dangerous to navigation

Marsh (swamp)   Submerged marsh

Wooded marsh   Mangrove

Woods or brushwood   Orchard

Vineyard   Scrub

Inundation area   Urban area

Relief

Relief symbols are used to show the characteristics of the land. Mountains, canyons, and other land features are shown by relief. The relief may be shown by contour lines or by more descriptive means such as technical shading methods. Contour lines are lines on a map that represent points of the same elevation. Contour lines are usually thin lines drawn freehand, with every fifth line being drawn thicker and broken somewhere to insert the elevation number. This number represents the elevation in feet above sea level of the line. Lettering on a map in conjunction with relief is usually done in vertical uppercase letters. Contour lines and other relief symbols are usually brown when the map is in color.

Water

Water features represent lakes, rivers, streams, and even intermittent waters. Water features are colored blue when the map is in color. Something unique about water features is that when labeled on a map, they are done using uppercase slanted letters. For example:

MISSISSIPPI RIVER

Look at Figure 5-1.

Vegetation

Vegetation features include forests, orchards, croplands, and other types of plant life. When a map is done in color, these features are usually green. Look at Figure 5-1. If the vegetation is planted by people, such as a field of corn, some maps classify these as culture.

Symbols have been made by various governmental agencies for their particular maps. Also, agencies of private industry may have their own mapping symbols. Some examples are:

Governmental

1. Federal Board of Surveys and Maps
2. National Oceanic and Atmospheric Administration (formerly U.S. Coast and Geodetic Survey)
3. U.S. Geological Survey
4. U.S. Forest Service
5. Army Map Service

Private Industry

1. American Railway Engineering Association
2. American Consulting Engineers Council

SPECIAL TECHNIQUES Other methods are often used to show symbols or to represent specific concepts on a map. For example, it was discussed earlier that relief is sometimes shown using sketching techniques. This is done to achieve special effects such as a three-dimensional appearance. You can see some examples in Figure 5-2.

You may even draw a representation of water features with rapids and whirlpools, depending on the purpose of your map (see Figure 5-3).

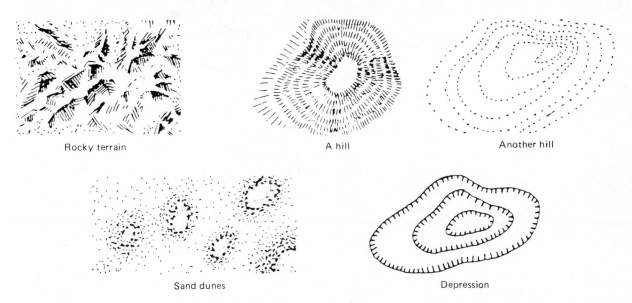

Rocky terrain   A hill   Another hill

Sand dunes   Depression

Figure 5-2   Topographical features.

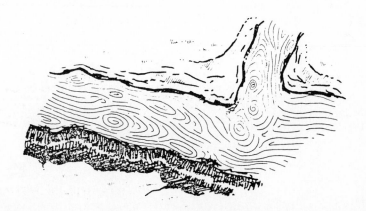

Figure 5-3   Water features using sketching methods.

Some maps may show specific crops grown in an area, or information about how land is being used. Some optional symbols are shown in Figure 5-4.

There are probably as many symbols as there are purposes for maps. You have seen a lot of standard symbols; there are many more that you may find on a map, and there also may be some minor differences from one map preparer to the next. Figure 5-5 displays some symbols.

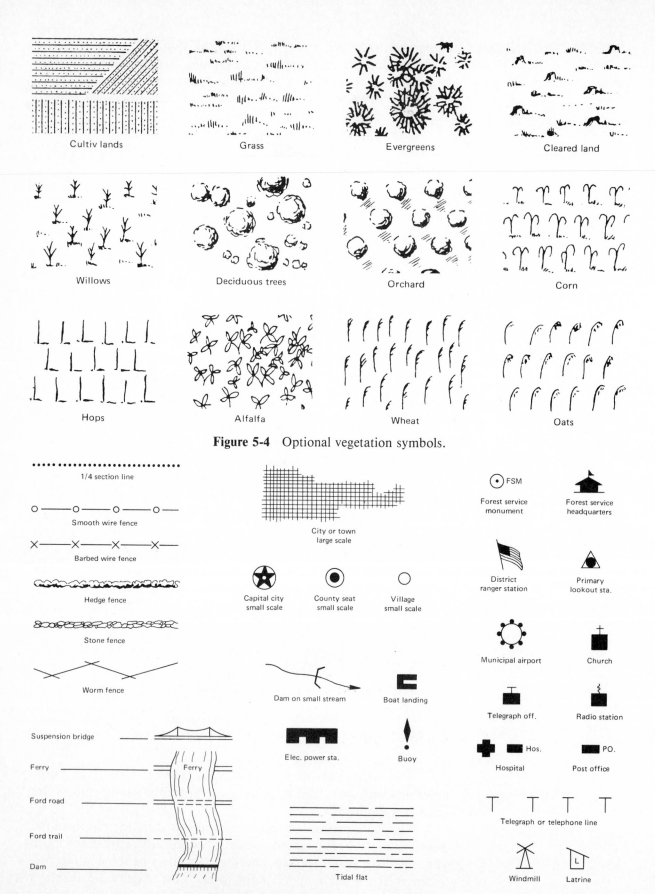

Figure 5-4 Optional vegetation symbols.

Cultiv lands

Grass

Evergreens

Cleared land

Willows

Deciduous trees

Orchard

Corn

Hops

Alfalfa

Wheat

Oats

1/4 section line

Smooth wire fence

Barbed wire fence

Hedge fence

Stone fence

Worm fence

Suspension bridge

Ferry

Ford road

Ford trail

Dam

Ferry

City or town
large scale

Capital city
small scale

County seat
small scale

Village
small scale

Dam on small stream

Boat landing

Elec. power sta.

Buoy

Tidal flat

FSM

Forest service
monument

Forest service
headquarters

District
ranger station

Primary
lookout sta.

Municipal airport

Church

Telegraph off.

Radio station

Hos.

Hospital

PO.

Post office

Telegraph or telephone line

Windmill

Latrine

Figure 5-5 More possible map symbols.

TEST

Part I

Define each of the following types of map symbols. Use your best freehand lettering.

5.1. Culture *REPRESENT WORKS OF PEOPLE*

5.2. Relief *USED TO SHOW CHARACTERISTICS OF THE LAND*

5.3. Water *REPRESENT LAKES, RIVERS, STREAMS AND INTERMITTENT WATERS*

5.4. Vegetation *SHOW FORESTS, ORCHARDS, CROPLANDS, AND OTHER TYPES OF PLANT LIFE*

Part II

In the space provided, carefully sketch an example of each of the following map symbols.

5.1. Hard-surface heavy-duty road

5.2. Hard-surface medium-duty road

5.3. Railroad, single track

5.4. Bridge and road

5.5. Building (dwelling)

5.6. Power transmission line

5.7. Perennial stream

5.8. Intermittent stream

5.9. Marsh

5.10. Orchard

Part III

Using shading techniques, sketch examples of each of the following terms.

5.1.   Rocky terrain

5.2.   Hill

5.3.   Depression

5.4.   Sand dune

5.5.   River

PROBLEM

P5.1. Draw a map in the space provided on Figure P5-1. Use the instructions here and on page 76 to construct the map.

    1. Use standard topographic symbols.

    2. Use normal drafting practices with pencil or ink as directed by your instructor.

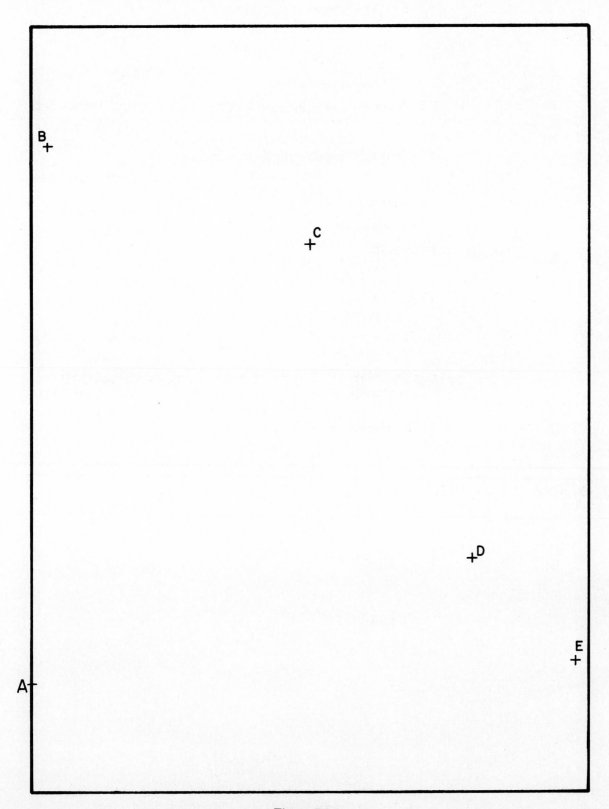

Figure P5-1

75

3. Use appropriate mapping colors.

4. North is toward the top of the page.

5. The scale is 1 in. = 200 ft.

6. From point *A*, draw a hard-surface medium-duty road with a bearing of N60°E.

7. Beginning 975′ from point *A*, draw a bridge over a 40-ft-wide river.

8. The center of the river runs through points *B, C, D,* and *E.* Use freehand lines to draw the river.

9. At 420 ft from point *A*, draw an improved light-duty road with a bearing of S26°E.

10. A small creek enters the river at point *C* with a bearing of 45°E for 385 ft, then turns with a radius of 200 ft and a bearing of N68°W.

11. North of the medium-duty road and southwest of the river, the entire area is orchard to within 200 ft of the river.

12. The area south of the medium-duty road and west of the light-duty road is a corn crop.

13. On both sides of the small creek is a grove of deciduous trees 100 ft wide.

14. All other areas are grasslands.

15. Two hundred feet northeast of the bridge and on the north side of the medium-duty road is a forest service headquarters.

6

Legal Descriptions and Plot Plans

This chapter shows the three typical ways of identifying property. Each individual property must be completely described by a survey. This survey becomes the legal description that keeps your land or property separate from your neighbor's property.

The topics covered include:

- Metes and bounds
- Lot and block
- Public land surveys
- Basic reference lines
- Rectangular system
- Plot plans
- Methods of sewage disposal

METES AND BOUNDS   Metes and bounds is a method of describing and locating property by measurements from a known starting point called a *monument*. Metes can be defined as being measurements of property lines expressed in units of feet, yards, or rods. Bounds are boundaries such as streams, streets, roads, or adjoining properties. The monument, or point of beginning of the system, is a fixed point such as a section corner, a rock, tree, or the intersection of streets.

The metes and bounds system is often used for describing irregularly shaped plats and is used as the primary method of describing plats in states east of the Mississippi River. A typical plat using metes and bounds is shown in Figure 6-1.

The following is a sample legal description using metes and bounds:

> BEGINNING at the intersection of the centerline of W. Powell Boulevard, formerly Powell Valley Road and the centerline of S.W. Cathey Road; thence running East along the centerline of W. Powell Boulevard 184 feet; thence South on a line parallel with S.W. Cathey Road, 200 feet; thence West on a line parallel with W. Powell Boulevard, 184 feet; thence North along the centerline of S.W. Cathey Road to the place of beginning; EXCEPTING therefrom, however, the rights of the public in and to that portion of the herein described property lying within the limits of W. Powell Boulevard and S. Cathey Road.

Figure 6-1 A typical plat using metes and bounds.

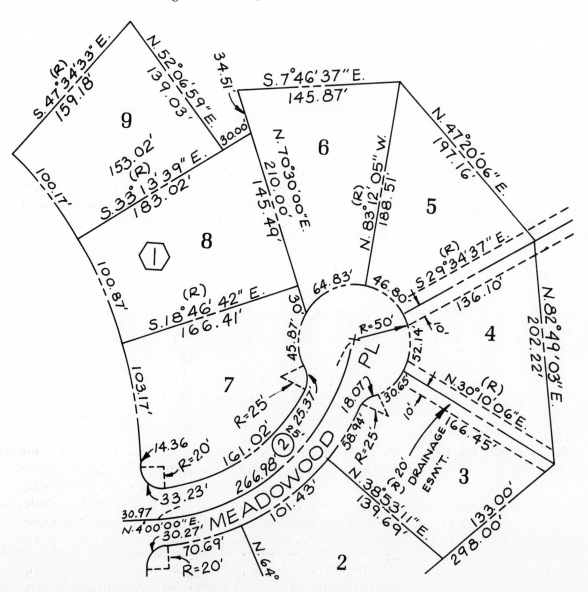

LOT AND BLOCK Lot and block is a method that describes land by referring to a recorded plat, the lot number, the county, and state. A legal lot and block must be filed with the county clerk or recorder as part of a plat, which is a map or plan of a subdivision. Look at Figure 6-2.

You will most often see the lot and block system used to describe small units of property in a subdivision. The exact boundaries of the subdivision may be described by the rectangular system or the metes and bounds system. Just remember that a necessary part of a plot plan, or plat, is the inclusion of an accurate legal description.

The following is an example of a lot and block description:

Lot 7, Block 135, Oregon City Subdivision, City of Oregon City, Clackamas County, State of Oregon.

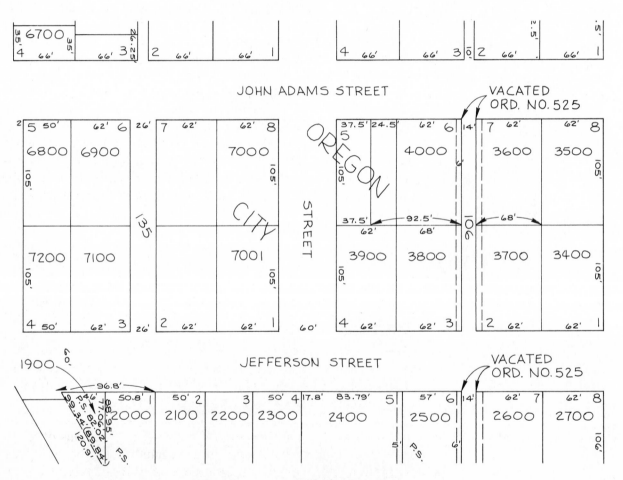

Figure 6-2 Lot and block.

RECTANGULAR SYSTEM Public Land Surveys

In the midwestern and far western states, or *public land states*, the U.S. Bureau of Land Management devised a *rectangular system* for describing land. The states involved in the public land surveys are:

Alabama, Alaska, Arizona, Arkansas, California, Colorado, Florida, Idaho, Illinois, Indiana, Iowa, Kansas, Louisiana, Michigan, Minnesota, Mississippi, Missouri, Montana, Nebraska, Nevada, New Mexico, North Dakota, Ohio, Oklahoma, Oregon, South Dakota, Utah, Washington, Wisconsin, and Wyoming. In 1850, the federal government bought 75 million acres from Texas, and these too are public lands. In the list above, you will note some southern, and even one southeastern state (Florida), included in the public land states. The public land states begin with Ohio. Its west boundary is the first principal meridian.

Basic Reference Lines

Each large portion of the public domain is a single *great survey*, and it takes on as much as it can use of one parallel of latitude and one meridian of longitude. The initial point of each great survey is where these two basic reference lines cross. This must be determined astronomically: a star-true point. The parallel is called the *base line*, and the meridian is called the *principal meridian*. There are 31 pairs or sets of these standard lines in the United States proper, and three in Alaska.

At the outset, each principal meridian was numbered. However, the numbering stopped with the sixth principal meridian, which passes Nebraska, Kansas, and Oklahoma. The balance of the 31 sets of standard lines took on local names. For example, in Oregon the public land surveys use the *Willamette meridian* for the principal meridian.

Rectangular-Township/Section System

Using base lines and meridians, an arrangement of rows of blocks, called *townships* is formed. Each township is 6 miles square. They are numbered by rows or tiers, the rows running east–west. These tiers are counted north and south from the base line. But instead of saying "tier one, tier two, tier three, . . . ," we say "township." For example, a township in the third tier north of the base line will be named, "Township No. 3 North," abbreviated as "T. 3N." Similarly, the third tier south of the base line will be named "T. 3S."

Townships are also numbered according to which vertical (north–south) column they are in. These vertical columns of townships are called *ranges*. Ranges take their numbers east or west of the principal meridian. A township in the second range east of a principal meridian is "R. 2E."

If we put these two devices of location together, the township in the third tier north of the base line and in the second range east of the principal meridian is "T. 3N., R. 2E." Figure 6-3 illustrates the arrangement of townships about the two reference lines. The tiers number as far north and south, and ranges number as far east and west, as that particular great public land survey goes.

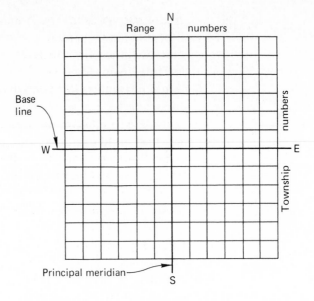

Figure 6-3   Arrangement of townships about the base line and principal meridian.

The Division of Townships

Figure 6-4 illustrates the division of a township into sections, and how the sections are numbered in each township.

A township is a square with sides approximately 6 miles long. At each mile along the 6-mile sides of a township, a line cuts across, forming a checkerboard of 36 squares, each a mile square. Each parcel of land, being approximately 1 mile square, contains 640 acres. These squares are *sections*. Sections are numbered 1 to 36, beginning at the northeast corner of the township and going across from right to left, then left to right, right to left, and so on, until all

Figure 6-4   Correct method of numbering sections within a township.

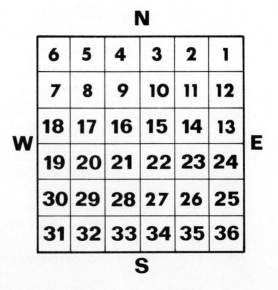

36 squares or sections are numbered. Remember, always start numbering with 1 in the upper right-hand corner, and 36 will always be in the lower right corner.

The Subdivision of Sections

Figure 6-5 illustrates some typical subdivisions of a section. The smallest subdivision shown in Figure 6-5 is 10 acres. However, even smaller subdivisions can be made, such as a residential subdivision of lots less than an acre in size.

Each section, as shown in Figure 6-5, may be subdivided into *quarter sections*. These are sometimes called *corners*: for example, "northwest corner of Section 16" or "NW 1/4 of Sec. 16."

A quarter section may be further divided into *quarter-quarters*; for example, "NW 1/4, NW 1/4 of Sec. 36." Further subdivisions of a quarter-quarter are illustrated in Figure 6-5.

In a written description of a portion of land, it will be noted that the smallest portion is written first, followed by the next larger portion, and so on. Looking at Figure 6-5 again, the smallest portions are two 10-acre subdivisions. Note that the description begins with the smallest and progresses to the largest: for example, "SE 1/4, SW 1/4, SW 1/4."

Figure 6-5 Some typical subdivisions of a section.

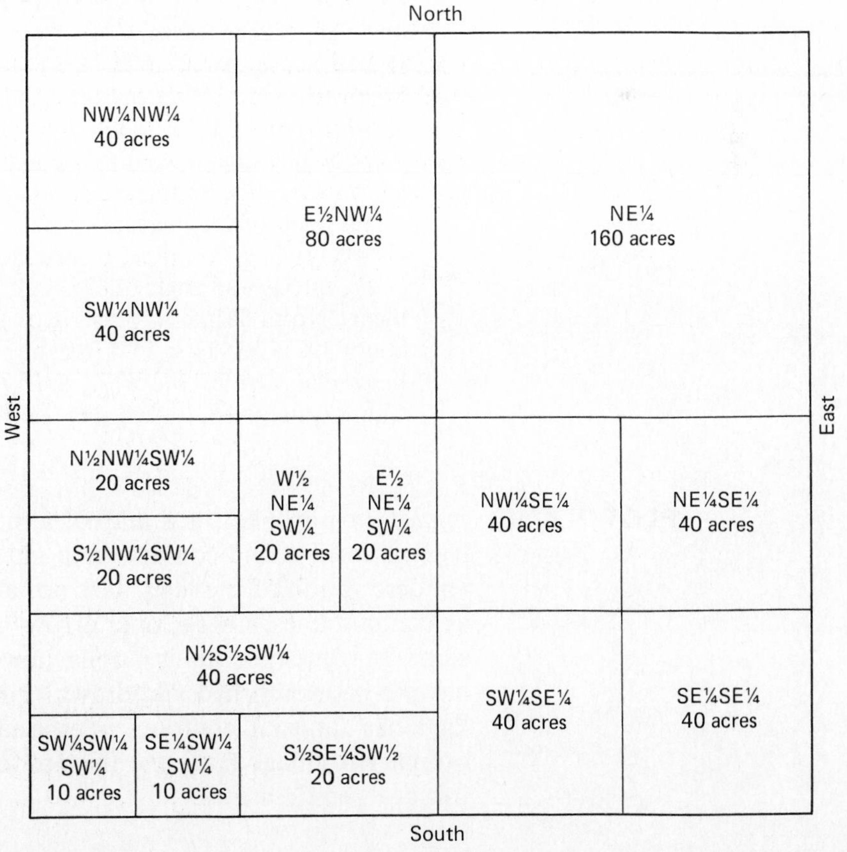

The Complete Description

The description of a small subdivision must include its relationship with the township, and then in turn relate to the reference lines of the great survey. Always remember that a complete description begins with the smallest division and progresses to the largest. An example of a legal description could look like this:

> The SW 1/4, SE 1/4 of Section 9 in Township 3 South, Range 2 East of the Willamette meridian, all being in Clackamas County, Oregon.

The following legal description reveals that a complete description of real property may include all three types of descriptions in combination. This example uses the rectangular system to identify the point of beginning, metes and bounds to describe the boundary lines, and a lot and block description that may also be used as an alternate description.

> Part of the Stephen Walker D.L.C. No. 52 in Sec. 13 T. 2 S. R. 1E., of the W.M., in the County of Clackamas and State of Oregon, described as follows:
>
> Beginning at the one-quarter corner on the North line of said Section 13; thence East 111.65 feet; thence South 659.17 feet; thence South 25°35′ East 195.00 feet to a ⅝-inch iron rod; thence South 56° West 110.0 feet to a ⅝-inch iron rod; thence South 34° East 120.79 feet to a ⅝-inch iron rod; thence South 19°30′ East 50.42 feet to a ⅝-inch iron rod; thence South 40°32′ East 169.38 feet to a ⅝-inch iron rod; thence South 49°29′ West 100.0 feet to a ⅝-inch iron rod; thence South 65°00′50″ West 51.89 feet to a ⅝-inch iron rod; thence South 52°21′30″ West 125.0 feet to a ½-inch iron pipe, being the true point of beginning; thence continuing South 52°21′30″ West 124.91 feet to a ½-inch iron pipe on the arc of a circle with a radius of 50.0 feet, the center point of which bears North 22°01′30″ West 50.0 feet from said last-mentioned iron pipe; thence Northeasterly along the arc of said circle, through an angle of 69°53′29″, 60.99 feet to a ½-inch iron pipe; thence North 88°05′01″ East 51.00 feet to a ½-inch iron pipe; thence South 39°55′35″ East 80.0 feet to the true point of beginning. ALSO known as *Lot 6, Block 3, CHATEAU RIVIERE,* in Clackamas County, Oregon.

PLOT PLANS   A *plat*, or plot plan, is a map of a piece of land. A plat becomes a legal document and contains an accurate drawing, as well as a written description of the land. It is not always necessary to show relief as contour lines. Often, only arrows are used to show direction of slope. In some specific situations, however, actual contour lines may have to be established and drawn. This may be necessary when the land has unusual contour, is especially steep, or has out-of-the-ordinary drainage patterns. Be sure to check with your jurisdiction for local requirements.

Requirements

Many items are necessary to make a plot plan a legal, working document. As a drafter, you can use the data as a checklist for proper completion of your plot. Be sure to check your local city or county regulations for any different requirements. Some of the following requirements may not apply to your plot plan. Many building departments require plot plans to be drawn on 8½ by 14-in. paper.

1. Legal description of the property.
2. Property lines, dimensions, and bearings.
3. Direction of north.
4. All roads, existing and proposed.
5. Driveways, patio slabs, parking areas, and walkways.
6. Proposed and existing structures.
7. Location of well and/or water service line. Location of wells on adjacent properties.
8. Location of proposed gas and power lines.
9. Location of septic tank, drainfield, drainfield replacement area, and/or sewer lines.
10. Dimensions and spacing of soil absorption field, or leach lines, if used.
11. Location of soil test holes, if used.
12. Proposed location of rain drains, footing drains, and method of disposal.
13. Ground elevation at lot corners, and street elevation at driveway centerline.
14. Slope of ground.
15. Proposed elevations of main floor, garage floor, and basement of crawl space.
16. Number of bedrooms proposed.
17. Proposed setback from all property lines.
18. Utility and drainage easements.
19. Natural drainage channels.
20. Total acreage.
21. Drawing scale: for example, 1 in. = 50 ft.

Figure 6-6 shows a typical plot plan. Notice that not all of the information is identified. Be sure to use all of the information that you need to describe your plot completely.

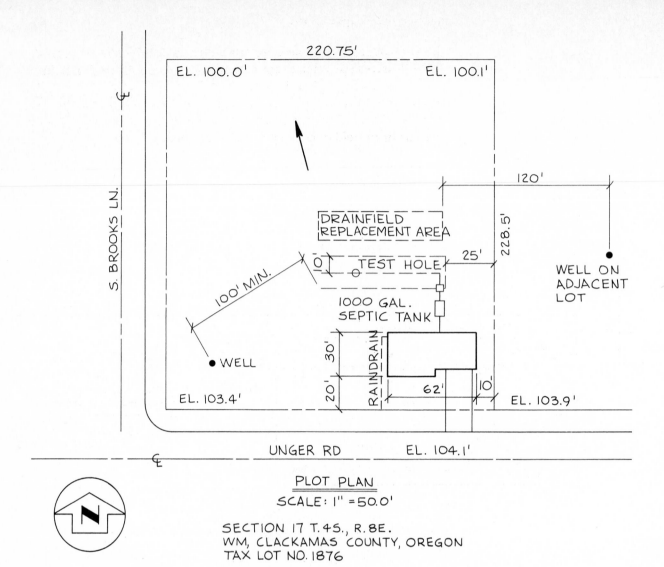

Figure 6-6 A typical plot.

METHODS OF SEWAGE DISPOSAL

Septic Tank

The conventional septic tank is usually a concrete or steel box where the wastewater from the house collects. Wastewater from toilets, bathtubs, showers, laundry, and kitchen are fed into this tank. It is designed to hold the water for 2 or 3 days, long enough for most of the heavy suspended material to sink to the bottom of the tank to form a sludge. Lighter, floating materials float to the top of the tank, where they remain trapped between the inlet and outlet pipes. After a couple of days, the wastewater portion leaves the tank as *effluent*. The effluent is discharged to the underground piping network, called a *soil absorption field, drainfield,* or *leach lines*.

The soil absorption field is an underground piping network buried in great trenches usually less than 2 ft below the surface of the ground. This field distributes the effluent over a large soil area, allowing it to percolate through the soil. The soil usually acts as an excellent filter and disinfectant by removing most of the pollutants and disease-causing viruses and bacteria found in the effluent. Figure 6-7 shows a section through a septic tank and a partial absorption field.

Now take a look at Figure 6-8 and you will see how the septic system should be drawn on a plot plan. Keep in mind that specific lengths of drainfield are determined by local requirements. Be sure that the drainfield lines run parallel to the contour lines.

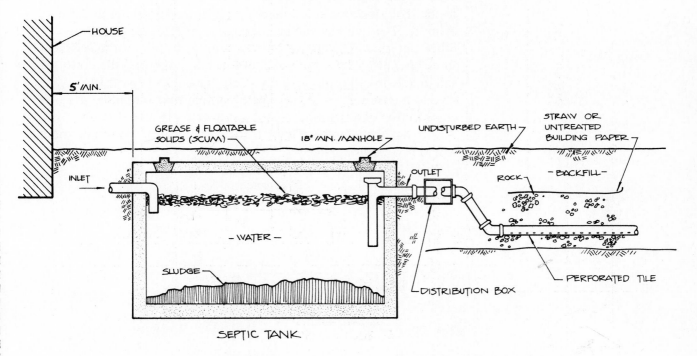

Figure 6-7 Cross section through a typical septic tank.

Figure 6-8 Sample plot showing a house and septic system.

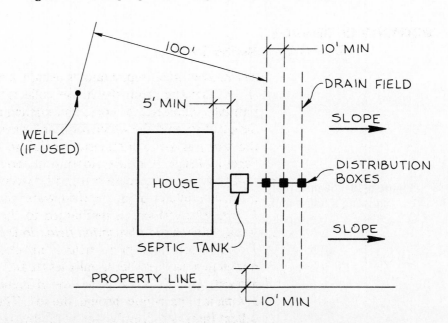

Cesspools

Cesspools have the same purpose as septic systems, that is, the breaking down and distribution of waste materials to an area of earth. The soil then acts as a filter to disperse pollutants. Cesspools are used in locations where the soil bearing strata is very porous. This would be an area of gravel or similar material of considerable depth. The structure that makes up the cesspool is a large concrete cylinder. This cylinder can be of precast concrete, concrete block, or other materials. The cesspool has slots at the bottom for the effluent to escape into a layer of gravel around the tank and then into the soil, made up of porous material. Your local soils department will be able to advise you as to the type of system that should be used. A cross section of a typical cesspool is shown in Figure 6-9.

Now take a look at Figure 6-10 and you can see how a cesspool will look on a plot plan.

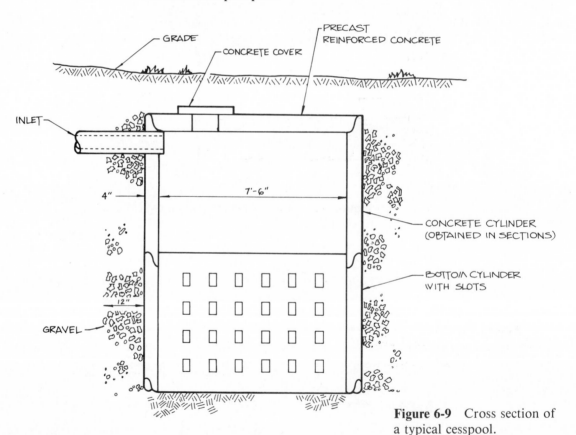

Figure 6-9 Cross section of a typical cesspool.

Figure 6-10 Sample plot showing a house and cesspool.

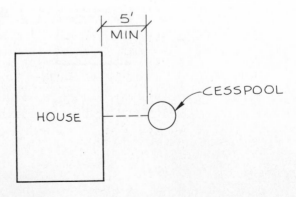

Public Sewers

In locations where public sewers are available, the plot plan should show a sewer line from the house to the public sewer, usually located in the street or in an easement provided somewhere near the property. This method of sewage disposal is often easier than the construction of a cesspool or septic system. See Figure 6-11 for a cross section of a sewer hookup.

Figure 6-12 shows how a conventional sewer hookup would be drawn on a plot plan.

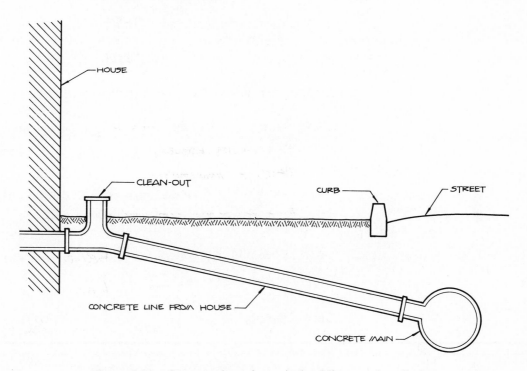

Figure 6-11   Cross section of a typical public sewer installation.

Figure 6-12   Sample plot showing a house and public sewer layout.

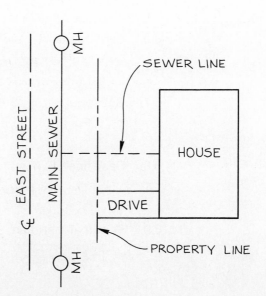

Utilities

The other utilities may be drawn into the property from the main lines that exist in a street or utility easement. These utilities may include electrical, gas, phone, and TV cable. Some utilities may be overhead, such as electrical. However, all utilities could be brought into the property from underground. Before you complete the plot plan, determine where the utilities will enter the property, so that they can be identified on the proposed plan.

TEST

Part I

Define the following terms using concise statements.

6.1. Metes and bounds *METHOD OF DESCRIBING AND LOCATING PROPERTY BY MEASURING FROM A KNOWN STARTING POINT CALLED A MONUMENT*

6.2. Lot and block *METHOD THAT DESCRIBES LAND BY REFERRING TO A RECORDED PLAT, THE LOT NUMBER, AND STATE.*

6.3. Township *FORMED USING BASELINES AND MERIDIANS, EACH TOWNSHIP IS 6 MI.²*

6.4. Section *1 MI² OR 640 & IT'S 1/36 OF A TOWNSHIP*

6.5. Plot plan *MAP OF A PIECE OF LAND*

6.6. Septic tank *A CONCRETE OR STEEL BOX WHERE WASTE WATER. FROM THE HOUSE COLLECTS. IT'S A BIG BOX FULL OF SHIT*

6.7. Cesspool *IT IS SIMILAR TO A SEPTIC TANK EXCEPT THAT THE EFFLUENT (SHIT)*

6.8. Base line *A PRINCIPAL PARALLEL USED IN ESTABLISHING THE RECTANGULAR SYSTEM OF LAND DESCRIPTION*

6.9. Principal meridian *ESTABLISHED AS A BASIS FOR ESTABLISHING A REFERENCE LINE FOR THE ORIGIN OF THE RECTANGULAR SYSTEM*

6.10.   Acre   *A PARCEL OF LAND MEASURE 43,316 FT².*

Part II

Given the section shown in Figure T6-1 with areas labeled by letters, provide the legal description and the number of acres for each area.

Figure T6-1

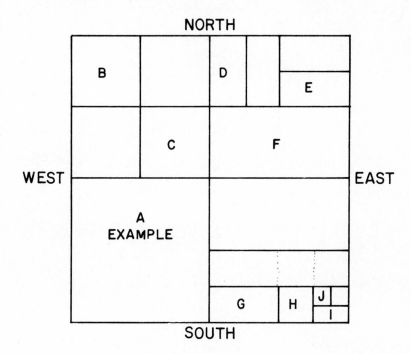

	Area	Legal Description	Acres
6.1	A	SW 1/4	160
6.2	B	NW ¼, NW ¼	40
6.3	C	SE ¼, NW ¼	40
6.4	D	W½, NW¼, NE¼	20
6.5	E	S½, NE¼, NE¼	20
6.6	F	S½, NE¼	80
6.7	G	S½, SW¼, SE¼	20
6.8	H	SW¼, SE¼, SE¼	10
6.9	I	S½, SE¼, SE¼, SE¼	5
6.10	J	NW¼, SE¼, SE¼, SE¼	2.5

Part III

Carefully sketch examples of plot plans that will display each of the following characteristics.

6.1. Public sewer to house on plot. Show driveway to house from street.

6.2. Cesspool to house on plot. Show driveway to house from street.

6.3. Septic system to house on plot. Show driveway to house from street.

PROBLEMS

P6.1. Given the rough sketch shown in Figure P6-1, draw a plot plan on 8½- by 14-in. vellum. The sketch is not to scale. Scale your drawing to fit the paper with a ½-in. minimum margin to the edge of the paper. Use the following information:

1. Use pencil, with freehand lettering and proper line technique.
2. You select the scale (e.g., 1 in. = 20 ft, 1 in. = 50 ft). The scale you select should make the plot plan as large as possible within the limits of the paper.
3. Label the plot plan, scale, legal description, elevations given, and north arrow.
4. Use as many of the plot plan requirements described in this chapter as possible.

Figure P6-1

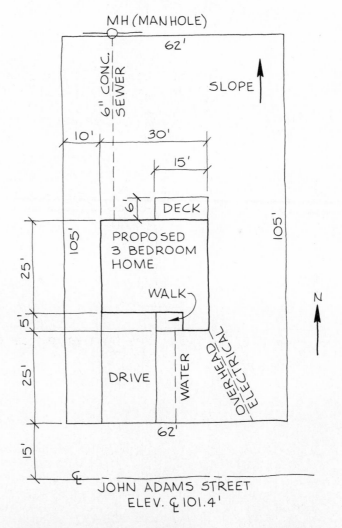

OREGON CITY ADDITION
LOT 7, BLOCK 2, OREGON CITY, OREGON
ELEV. MAIN FL. 101.8'
ELEV. BSMT. FL. 92.8'

P6.2. Given the rough sketch in Figure P6-2, draw a plot plan on 18- by 24-in. or 17- by 22-in. vellum. The sketch is not to scale. Scale your drawing for best utilization of the paper size. Use the following information:

1. Use pencil, with freehand lettering, and proper line technique.
2. Construct your own title block with title, legal description, your name, the scale, and a north arrow.
3. Use as many of the plot plan requirements described in this chapter as possible.

Figure P6-2

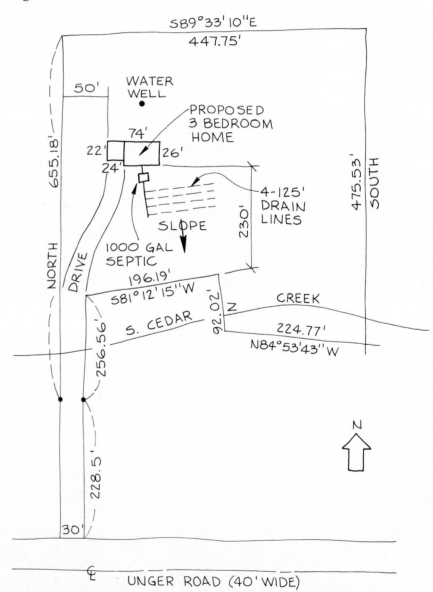

7

Contour Lines

The topography of a region is best represented by contour lines. The word topography comes from the Greek words *topos*, a place, and *graphein*, to draw. The most common method of "drawing a place" for mapping purposes is to represent differences in elevation with contour lines. These lines connect points of equal elevation. They can also reveal the general lay of the land and describe certain geographical features to those trained in topographical interpretation.

CONTOUR LINE CHARACTERISTICS

The best example of a contour line is the shore of a lake or reservoir. The water level represents one contour line because the level of the lake is the same in all places. By late summer, many reservoirs are lowered considerably and previous water levels are seen as lines, contour lines. The space between these lines is termed the *contour interval*.

If you observe the contour lines of a reservoir closely, you can see that they do not touch, and run parallel to each other. One line can be followed all the way around the reservoir until it closes on itself. This basic characteristic is shown in Figure 7-1. Note also that the contour lines are generally parallel, and they never cross one another.

Slopes

The steepness of a slope can be determined by the spacing of the contour lines. A gentle slope is indicated by greater intervals between the

Figure 7-1 Contour lines formed by lapping water at different levels in a reservoir. (Courtesy City of Portland, Oregon)

contours (see Figure 7-2), whereas a steep slope is evident by closely spaced contours (see Figure 7-3).

Slopes are not always uniform. A *concave* slope flattens toward the bottom, as seen in Figure 7-4. The steepness of the upper part of the slope is represented by the close contour lines. A *convex* slope is just the opposite and develops a steeper gradient as it progresses. The contour lines are closer together near the bottom of the slope, as seen in Figure 7-5.

An exception exists to the rule that contours never touch. In Figure 7-6, the contours are obviously touching. Can you determine the type of landform shown in Figure 7-6? It is a cliff. Sheer cliffs and vertical rock walls of canyons are easy to spot on topographical maps because of the sudden convergence of contour lines.

Streams and Ridges

Those of you who have taken hikes in the mountains know that as a level trail traverses a hill, you find yourself walking in a large "U"

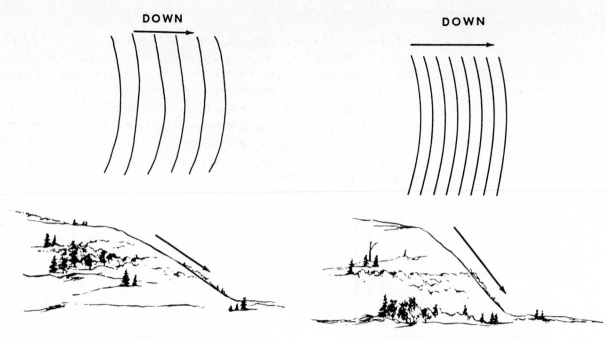

Figure 7-2   Uniform gentle slope.

Figure 7-3   Uniform steep slope.

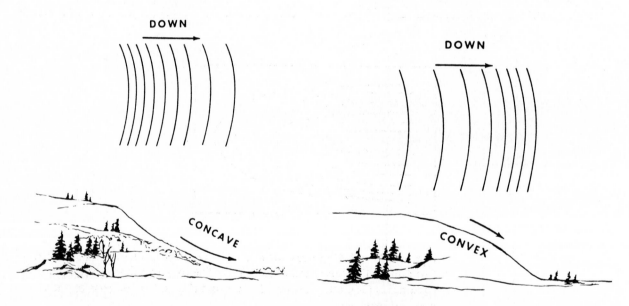

Figure 7-4   Concave slope.

Figure 7-5   Convex slope.

Figure 7-6   Contours merge to form a cliff.

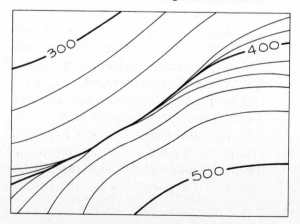

or horseshoe. As you approach a stream between two hills, the trail begins to form a "V" pointing upstream, as shown in Figure 7-7a. Near stream junctions, an "M" is often formed, as shown in Figure 7-7b. The peaks of the M point upstream.

The characteristics shown in Figure 7-7 can quickly reveal slope and stream flow and slope directions to an experienced map reader. Notice that the bottom of the V points upstream. Contour lines form a U around a hill or ridge and the bottom of the U points downhill, as shown in Figure 7-7c.

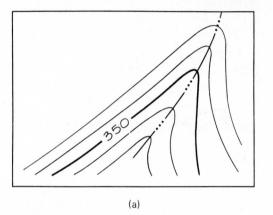

(a)

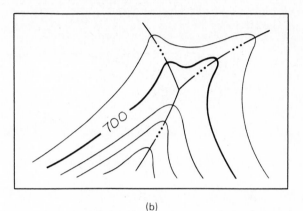

(b)

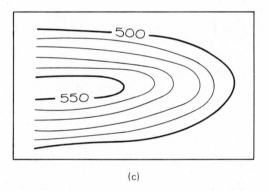

(c)

Figure 7-7 (a) Contours form a "V" pointing upstream. (b) Contours form an "M" above stream junctions. The tops of the M point upstream. (c) Contours form a "U" around ridges. The bottom of the U points downridge.

Relief Features

The features shown in Figure 7-8 are easily spotted on contour maps. The peak of a hill or mountain in Figure 7-8a is a common feature and may be accompanied by an elevation. Here, the contour lines form circles of ever-decreasing diameter. Two high points or peaks side by side form a *saddle*, such as that shown in Figure 7-8b. A saddle is a low spot between two peaks.

The special contour line in Figure 7-8c is used to represent a depression in the land. The contour line has short lines pointing into the depression, and is used to identify human-made features such as

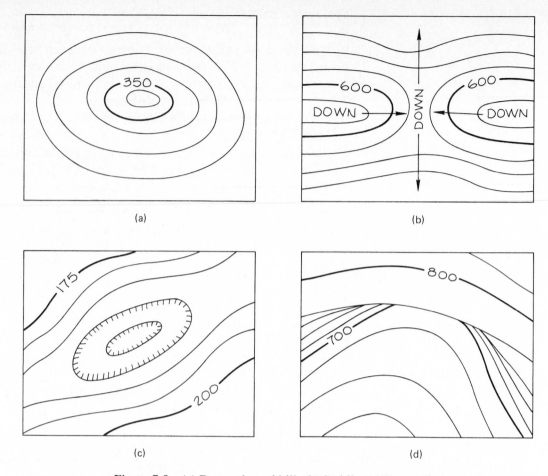

Figure 7-8 (a) Dome-shaped hill. (b) Saddle. (c) Depression. (d) Overhang.

quarries and pits, or natural features such as the limestone sinkholes common in the southeastern United States.

The feature seen in Figure 7-8d is similar to a cliff, but the lower contour lines actually seem to go under the higher ones. In reality they do, and portray an overhang, a feature found in rocky, highly sculpted and mountainous terrain.

TYPES OF CONTOUR LINES

Index Contours

Every fifth line on a topograhic map is an index contour. This aids the map reader in finding references and even-numbered elevations. The index contour is normally a thick line and is broken at intervals and labeled with its elevation (see Figure 7-9).

Intermediate Contours

The remaining contours in Figure 7-9 are intermediate contours and represent the intervals of elevation between the index lines. There are four of these lines between index contours. These lines are not normally labeled, but can be if the scale and function of the map dictate.

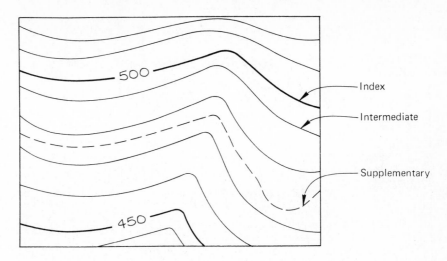

Figure 7-9 Types of contour lines.

Supplementary Contours

The supplementary contour is not as common as the others. It is used when the normal contour interval is too large to illustrate significant topographic features clearly. They are usually given the value of half the contour interval (see Figure 7-9).

PLOTTING CONTOUR LINES FROM FIELD NOTES

As a rule, topgraphic maps tend to be somewhat inaccurate in representing the true shape of the land. General features and large landforms can be shown accurately, but small local relief is often eliminated. This is especially true of contour maps created from aerial photographs. Trees and vegetation may hide the true shape of the land.

A more accurate view of the land can be obtained by ground survey. This process of mapping establishes spot elevations from known points. Field notes are then plotted as contour lines on the new map. Let us take a closer look at this process.

Control Point Survey

Plots of land may often be surveyed in such a matter that contour maps can be created from the survey or field notes. The creation of these maps depends on a certain amount of interpolation (guesswork), but a good knowledge of landforms, slopes, roads, and stream characteristics may ensure a greater degree of success in the mapping operation.

The control point survey is a common method of establishing elevations for use in contour mapping. Figure 7-10 is an example of the two basic steps in the mapping process. The surveyor's field notes (elevations) are plotted on the map and labeled. Next a contour interval is chosen. This depends on the purpose of the map and the elevation differential within the plot. Contour lines are then drawn to connect equal elevation points. Notice in Figure 7-10 that

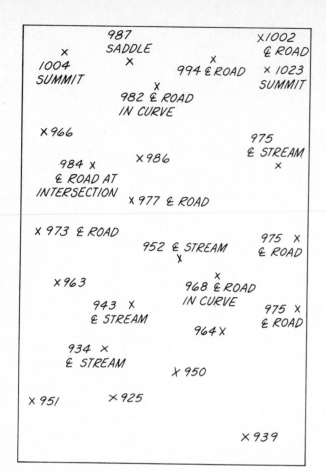
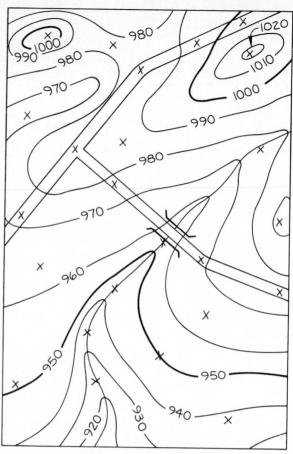

Figure 7-10 Contour map plotted using control point survey.

prominent features on the landscape are surveyed. These are the control points. Their elevations may not be an even number. In this case, the drafter must "interpolate" between two control points to find the even number. While we are on the subject, let us discuss interpolating a little further.

Interpolating Contour Lines

To *interpolate* means to insert missing values between numbers that are given. This is often the situation that the drafter faces when plotting field notes. Inserting missing values may seem like guesswork, but it can be accomplished with a certain amount of accuracy if five basic steps are followed.

Step 1: Establish elevations at stream junctions.

Given the elevations at points A and B in Figure 7-11a, we must determine the elevation of the stream junction at C. With dividers or scale, we find that C is two-thirds of the distance from A to B. The elevation difference between A and B is 60 ft. As ⅔ of 60 is 40, the elevation of point C is 170 ft. Using this method, determine the elevation of the stream junction at F.

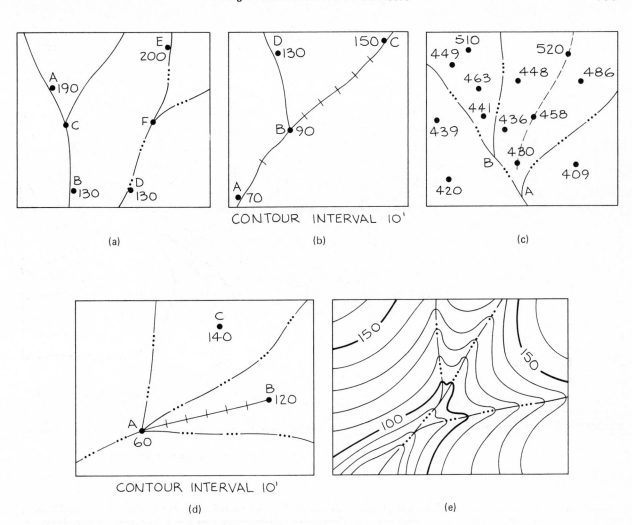

Figure 7-11 Interpolating contour lines.

Step 2: Locate points where contours cross streams.

Given a contour interval of 10 ft in Figure 7-11b and spot elevations at points *A, B, C,* and *D,* we must locate stream crossings of all contour lines. The vertical elevation between points *A* and *B* is 20 ft. Divide the distance between *A* and *B* in half and the 80-ft contour can be located at that point. The vertical distance between points *B* and *C* is 60 ft. The line between these two points must be divided into six equal segments. The first new point above *B* is 100 ft in elevation. Determine the stream crossings of the contour lines between points *B* and *D.*

Step 3: Locate ridges as light construction lines.

Areas of higher elevation usually separate streams except in swamps and marshes. To locate these ridges or *interfluves* (Figure 7-11c), sketch a light dashed line beginning at the stream junction and connecting the highest elevation points between the two streams. Sketch a similar line beginning at stream junction *B.*

Step 4: Determine the points on the ridge lines where contours cross.

This step is performed in exactly the same manner as step 2. In Figure 7-11d, the space between point *A* and the spot elevation of 120 ft is 60 ft. The line is divided into six equal parts as in step 2. Locate the contour crossings between points *A* and *C*.

Step 5: Connect the points of equal elevation with contour lines.

As you do this, keep in mind the characteristics of contour lines discussed previously (see Figure 7-11e). Also be aware that interpolating distances is necessary in situations other than those discussed in the five steps. Keep in mind that by interpolating contour lines, we are using an assumption termed *uniform slopes*. This is based on equal spacing of contours between known points. Some do's and don'ts are illustrated in Figure 7-12.

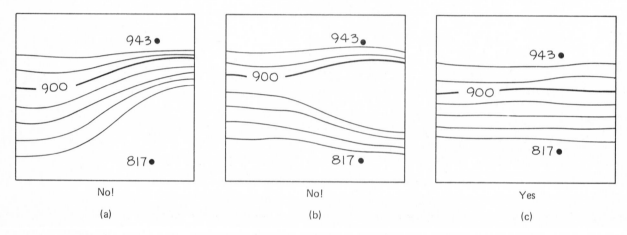

No!	No!	Yes
(a)	(b)	(c)

Figure 7-12 Interpolating contour lines using the "uniform slopes" concept.

Grid or Checkerboard Surveys

Using a grid, the surveyor divides the plot of land into a checkerboard. The drafter plots this grid as shown in Figure 7-13a. The size of the squares is determined by the surveyor based on the land area, topography, and elevation differential. The vertical lines of the grid in Figure 7-13 are labeled with letters and are 20 ft apart. The horizontal lines are labeled 0 + 00, 0 + 20, 0 + 40, 0 + 60, and so on, and are called *stations*. The stations are also 20 ft apart.

The elevation of each grid intersection is recorded in the field notes shown in Table 7-1. Using the surveyor's field notes, we see that grid point A–0 + 40 has an elevation of 100.0 ft. What is the elevation of point C–0 + 60?

To begin plotting the elevations, draw a grid at the required scale, or use grid paper. Next, using the field notes, label all the grid

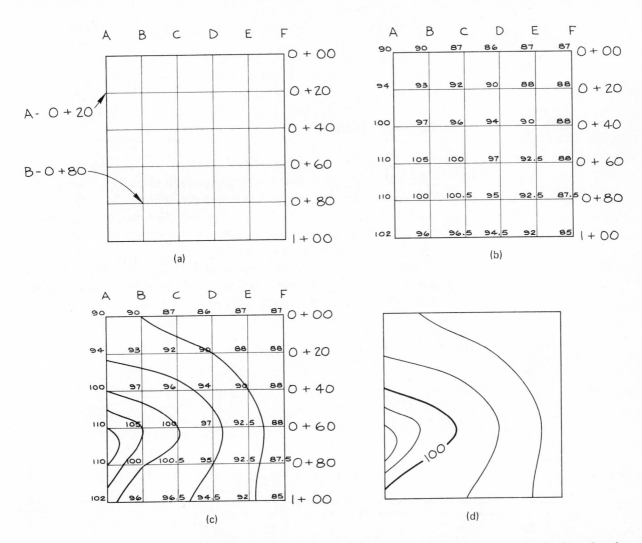

Figure 7-13 (a) For a grid survey, land is divided into a checkerboard and labeled. (b) All grid intersections are labeled. (c) Connect the elevations with freehand lines. (d) Completed contour map of grid survey.

Table 7-1 Grid Survey Field Notes

Station	*Elev.*	*Station*	*Elev.*	*Station*	*Elev.*
A–0 + 00	90.0	C–0 + 00	87.0	E–0 + 00	87.0
A–0 + 20	94.0	C–0 + 20	92.0	E–0 + 20	88.0
A–0 + 40	100.0	C–0 + 40	96.0	E–0 + 40	90.0
A–0 + 60	110.0	C–0 + 60	100.0	E–0 + 60	92.5
A–0 + 80	110.0	C–0 + 80	100.5	E–0 + 80	92.5
A–1 + 00	102.00	C–1 + 00	96.5	E–1 + 00	92.0
B–0 + 00	90.0	D–0 + 00	86.5	F–0 + 00	87.0
B–0 + 20	93.0	D–0 + 20	90.0	F–0 + 20	88.0
B–0 + 40	97.0	D–0 + 40	94.0	F–0 + 40	88.0
B–0 + 60	105.0	D–0 + 60	97.0	F–0 + 60	88.0
B–0 + 80	100.0	D–0 + 80	95.0	F–0 + 80	87.5
B–1 + 00	96.0	D–1 + 00	94.5	F–1 + 00	85.0

intersections with elevations (see Figure 7-13b). Connect the elevations with freehand lines. The drafter is aware of the required contour interval at this point and must decide which elevation points to connect and when to interpolate between uneven elevation points (Figure 7-13c). Figure 7-13d shows the finished contour map.

Contour Labeling

Most topographic maps show written elevations only on index contour lines. These labels are normally enough to give the map reader sufficient reference lines to work with. Figure 7-14 shows the method in which labeling is done. The elevation numbers are placed so that they are not upside down. Contour line labels are located at regular intervals along the contour line.

Property plats, highway maps, and special maps of many kinds may require every contour to be labeled. In this case the drafter must use good layout and spacing techniques to achieve a balanced and uncluttered appearance.

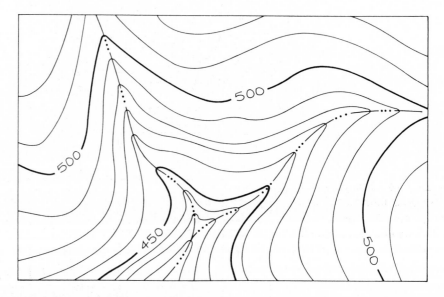

Figure 7-14   Contour line labeling.

ENLARGING CONTOUR MAPS

The civil drafter may be required to enlarge contour maps to show greater cultural detail or to define the topography with additional contour lines. Other than the photographic process or the use of enlarging projectors, the grid system is the best drafting method to use when enlarging (or reducing) a map.

Grid Layout

First, a grid must be drawn on the existing map. The size of the grid squares depends on the complexity of the map and the amount of detail you wish to show. You may be instructed to enlarge the map to twice its present size or a scale for the new map will be specified. If you draw the grid on the existing map with ¼-in. squares and wish to double the size, the grid for your new map is drawn with ½-in. squares. Suppose that the scale of the original map is 1 in. = 1000 ft and the new map is to be 1 in. = 250 ft. If you draw the grid on the original using ¼-in. squares, the new grid is then drawn using 1-in. squares—four times the size.

The map in Figure 7-15 must be enlarged to twice its size and has been overlayed with a ¼-in. grid. Notice, too, that the vertical and horizontal grids have been numbered to avoid confusion when transferring points. Keep in mind that when a linear measurement is doubled, the area of that map is increased four times the original size. The appearance of Figure 7-16 illustrates this point.

Figure 7-15 Grid drawn over original map.

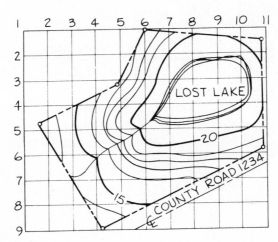

Figure 7-16 Enlarged map uses same number of grid lines, but dimensions of squares are doubled.

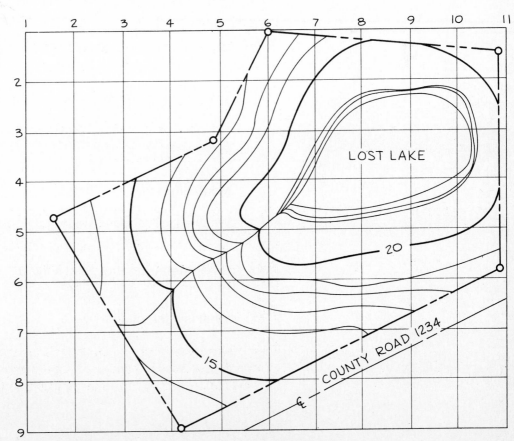

The grid for the new map is constructed using the same number of squares as the original and is labeled the same. The only difference is its size (see Figure 7-16). The squares now measure ½ in.

Map Construction

Map features can now be transferred from the original to the enlargement by eye, engineer's scale, or proportional dividers. Measurements provide the greatest degree of accuracy, but the drafter soon realizes that "eyeballing" features on or near grid intersections is accurate enough for the purposes of the map.

The two grids that you draw should be exactly the same, as well as the labeling you use to identify the vertical and horizontal grid lines. Compare Figures 7-15 and 7-16. The process of transferring from the original to the enlargement is relatively simple. Choose one square and measure or estimate where features touch or cross the grid lines of that square. Transfer that measurement to the same square on the enlarged grid, remembering, of course, to increase its size proportionately.

The grid enlargement method can produce accurate maps provided that the drafter establishes a proper coordinate system, uses good measuring techniques, and avoids unnecessary "artistic license."

TEST

7.1.   What do contour lines represent? *Connect points of equal elevation*

7.2.   List four characteristics of contour lines.

7.3.   Sketch the following features using contour lines:

Mountain peak   Saddle   Depression

7.4.   What is the function of index contours?

7.5.   Briefly define control point survey.

7.6.   What is interpolation?

7.7.   What type of survey divides the land into a checkerboard?

7.8.   Explain the grid system of map enlarging.

7.9.   What letter of the alphabet is formed when a contour line crosses a stream?_____ In which direction does this letter point?_____

7.10.   What letter is formed when a contour line wraps around a hill?_____ In which direction does the letter point?_____

PROBLEMS

P7.1.   Using proper methods of contour line interpolation, establish contour lines for the problems given in Figure P7-1. Use indicated contour intervals. Label all index contours and keep in mind the characteristics of contours discussed in this chapter.

Figure P7-1

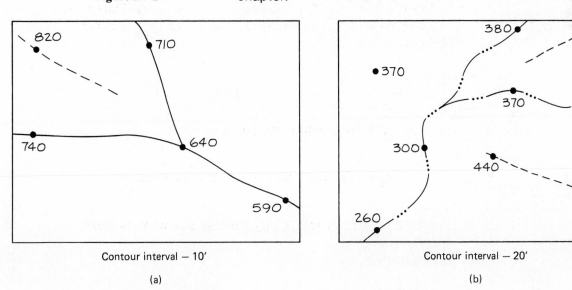

Contour interval — 10'

(a)

Contour interval — 20'

(b)

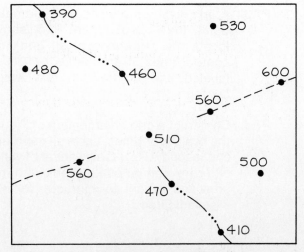

Contour interval — 10'

(c)

P7.2. Given the surveyor's field notes from a control point survey, establish contour lines at the contour interval indicated in Figure P7.2. Index contours should be a heavy line weight and should be labeled. Use interpolation where required.

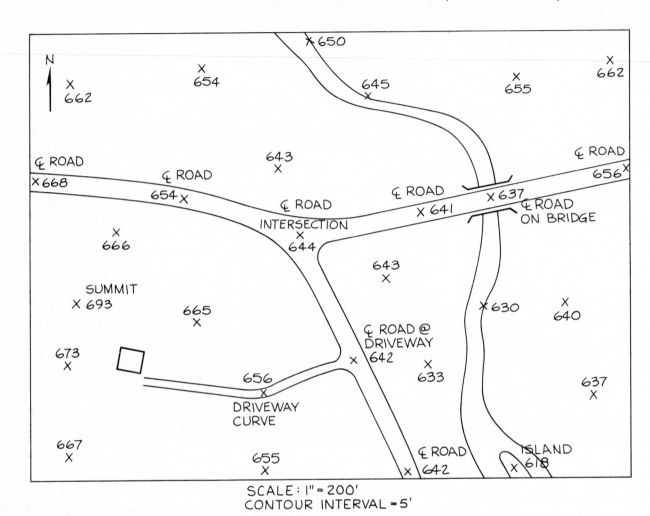

SCALE: 1" = 200'
CONTOUR INTERVAL = 5'

Figure P7-2

P7.3. A grid survey produced the field notes given in Table P7-1. Using the grid of Figure P7-3 (shown on page 112), first locate the elevations of all grid intersections. Then plot contour lines at an interval of 10 ft. Index lines should be labeled and contrast in line weight with intermediate contours.

P7.4. Construct a grid enlargement of Figure P7-2 or P7-3. Increase the scale four times. Label all contours. Include a north arrow and legend. Use a C-size sheet of vellum or Mylar. Either pencil or ink can be used (check with instructor). Make a diazo print and submit to instructor for evaluation, unless otherwise indicated.

Table P7-1 Grid Survey Field Notes

Station	Elev.	Station	Elev.	Station	Elev.
A–0 + 00	592	D–0 + 00	577	G–0 + 00	602
A–0 + 20	595	D–0 + 20	536	G–0 + 20	592
A–0 + 40	599	D–0 + 40	531	G–0 + 40	561
A–0 + 60	583	D–0 + 60	519	G–0 + 60	529
A–0 + 80	560	D–0 + 80	468	G–0 + 80	460
A–1 + 00	558	D–1 + 00	475	G–1 + 00	380
A–1 + 20	577	D–1 + 20	492	G–1 + 20	395
A–1 + 40	589	D–1 + 40	496	G–1 + 40	422
A–1 + 60	594	D–1 + 60	498	G–1 + 60	437
B–0 + 00	587	E–0 + 00	579	H–0 + 00	584
B–0 + 20	600	E–0 + 20	562	H–0 + 20	568
B–0 + 40	648	E–0 + 40	535	H–0 + 40	536
B–0 + 60	594	E–0 + 60	507	H–0 + 60	507
B–0 + 80	537	E–0 + 80	450	H–0 + 80	441
B–1 + 00	543	E–1 + 00	437	H–1 + 00	381
B–1 + 20	561	E–1 + 20	463	H–1 + 20	372
B–1 + 40	563	E–1 + 40	465	H–1 + 40	406
B–1 + 60	565	E–1 + 60	464	H–1 + 60	427
C–0 + 00	571	F–0 + 00	602	I–0 + 00	555
C–0 + 20	576	F–0 + 20	586	I–0 + 20	537
C–0 + 40	563	F–0 + 40	560	I–0 + 40	513
C–0 + 60	578	F–0 + 60	532	I–0 + 60	483
C–0 + 80	500	F–0 + 80	461	I–0 + 80	442
C–1 + 00	518	F–1 + 00	394	I–1 + 00	382
C–1 + 20	536	F–1 + 20	428	I–1 + 20	359
C–1 + 40	535	F–1 + 40	444	I–1 + 40	391
C–1 + 60	534	F–1 + 60	451	I–1 + 60	417

P7.5.   Identify the features shown below.

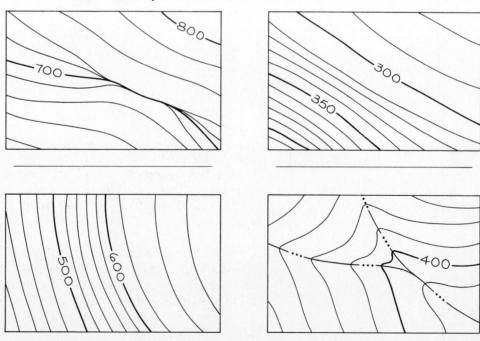

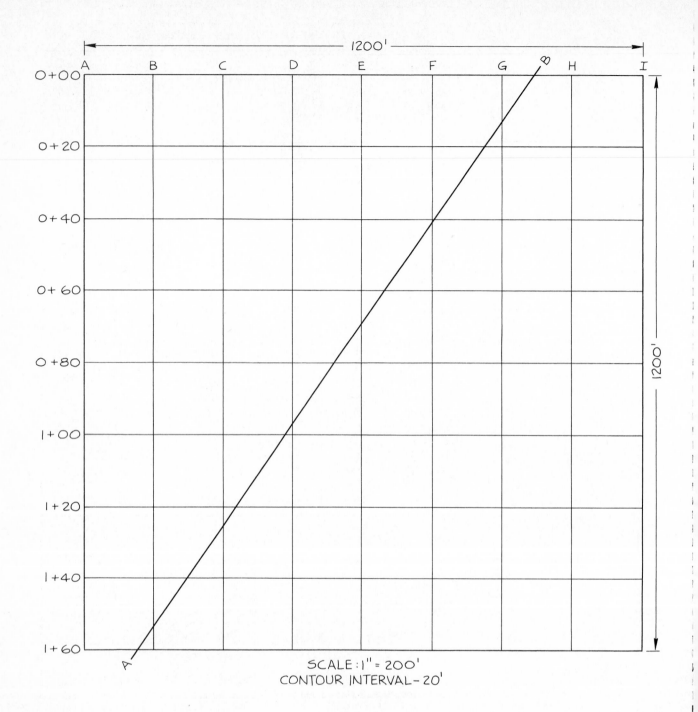

SCALE : 1" = 200'
CONTOUR INTERVAL - 20'

Figure P7-3

112

8

Profiles

A *profile* is an outline. An *artistic profile* is the outline of a face from the side and a *map profile* is the outline of a cross section of the earth. Profiles are drawn using the information given on contour maps. Their uses include road grade layout, cut-and-fill calculations, pipeline layouts, site excavations, and dam and reservoir layout. This chapter examines basic profile construction from contour maps and the *plan and profile* commonly used by civil engineering firms for underground utility location and layout, and highway designs.

CONTOUR MAP PROFILES

Map Layout

A straight line should be drawn on the map where the profile is to be made, as shown in Figure 8-1. The line between points *A* and *B* may be a proposed road or sewer line, and it may be at an angle other than horizontal on your drawing board. For ease of projection, turn the map so that the profile line is horizontal and aligned with the horizontal scale of your drafting machine. A clean sheet of paper can then be placed directly below the profile line and used to construct the cross section (see Figure 8-2).

Profile Construction

The horizontal scale of the profile is always the same as the map because the profile is projected from the map. The vertical scale may

114

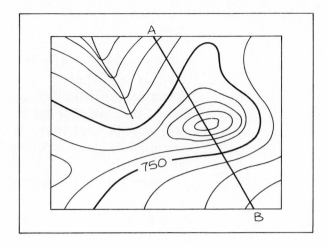

Figure 8-1 Profile to be cut along line *AB*.

Figure 8-2 Correct relationship of map, drawing paper, and drafting machine for profile construction.

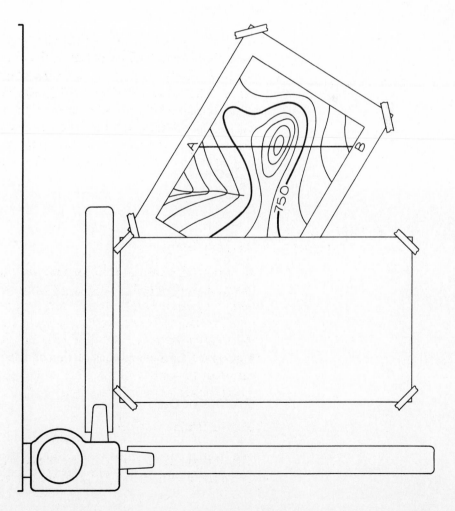

be exaggerated to give a clear picture of the shape of the land. The amount of exaggeration depends on the relief of the map, the scale of the map, and the purpose of the profile.

The length of the profile is established by projecting end points *A* and *B* to your paper. The height of your profile depends first on the amount of relief between points *A* and *B*. Find the lowest and highest contours that line *AB* crosses and subtract to determine the total amount of elevation to be shown in the profile. This enables you to establish a vertical scale to fit the paper and best show the relief.

When constructing the vertical profile scale, provide an additional contour interval above and below the extreme points of the profile. Also notice in Figure 8-3 that the vertical scale is labeled along one side and the scales are written below the profile. The vertical scale is sometimes written vertically near the scale values.

Project horizontal lines from the vertical scale values across the drawing, then project all points from the map where the profile line crosses contour lines. Notice in Figure 8-3 that a point on the profile is established where vertical and horizontal lines of the same elevation intersect. Once all these points are established, connect them with a freehand line. A sectioning symbol or shading is used to indicate the ground that is cross-sectioned.

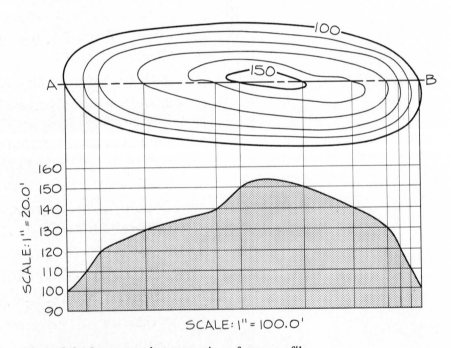

Figure 8-3 Layout and construction of map profile.

Profiles of Curved Lines

Layout of a profile from a straight line is simple, but plotting a profile from a curved line involves an additional step. Before the profile can be drawn, curved line *AB* must be established along a related straight line, *AB'*. This can be done with dividers, compass, or engineer's scale (see Figure 8-4). Label the new points on line *AB'* to avoid any confusion. Note that the distance 0–1 is transferred to 0–1', 1–2 to 1'–2' and so on.

From this point, creating the profile is the same as it is for a straight line. Be certain that you project the actual contour crossings from point *A* to 0 and the newly established points, 1' to 4', from 0 to *B'*. See Figure 8-5 for the proper method of projection.

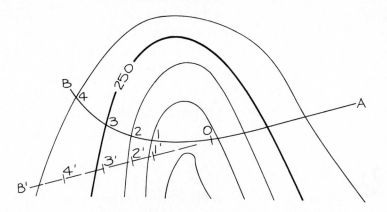

Figure 8-4 Profile construction of a curved line. Measure distances on *AB* and establish them on straight line *AB'*.

Figure 8-5 Construct profile of curved line *AB* from new straight line *AB'*.

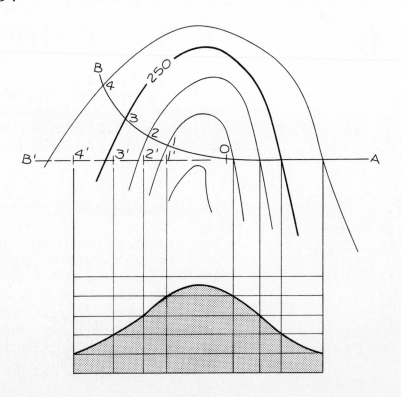

PLAN AND PROFILE   Description and Uses

The plan and profile as used by civil engineers and state highway departments can be compared to a top and front view in mechanical drafting, where the front view is a section. The drawing is convenient to use, for it allows both the plan and cross section of a specific area to be shown on the same sheet. The plan is always placed above the profile. The uses of this type of map are many. Transportation departments use the plan and profile extensively for layout and design of roads and transit systems as seen in Figure 8-6. The profile view is often done along the centerline of a road to illustrate gradient and curves. Civil engineering firms employ the plan and profile when designing subdivision street layouts and underground utility locations (see Figure 8-7).

Layout and Construction

The scales most commonly used for the plan view are 1 in. = 100 ft and 1 in. = 50 ft. All pertinent information necessary for the map is drawn on the plan. The plan is normally long and narrow because it is illustrating linear features, such as roads and sewer lines. Refer to Figure 8-7 for proper layout of the plan and profile. Before locating the plan on your drawing paper, it is important to know how much vertical elevation is to be shown in the profile.

The vertical scale of the profile is normally 1 in. = 10 ft when the plan scale is 1 in. = 100 ft, and 1 in. = 5 ft when the plan scale is 1 in. = 50 ft. Using the appropriate scale, determine from the field notes the amount of elevation to be shown in the profile. This information allows the drafter to decide what size of paper to use and where to locate the views on the paper. Because the profile is projected from the plan, the horizontal scale of the two views are the same.

Remember to allow some space beyond the highest and lowest elevations of the profile. Construct a grid of horizontal lines at even elevation points and vertical lines at even station points. Using this grid, the drafter can accurately locate manhole stations, grade, and sewer pipe elevations.

Terms and Symbols

The drafter should keep in mind that certain standard symbols and terms exist that are used frequently throughout the industry. But standards are sometimes modified and each company may alter things to suit their needs. With this in mind, the drafter should always be aware of the company standards in use at the time.

A common term used by everyone in civil drafting and surveying is *station point*. Station points are established by surveyors every 100 ft. The first station point is 00 + 00, then 01 + 00, 02 + 00, and

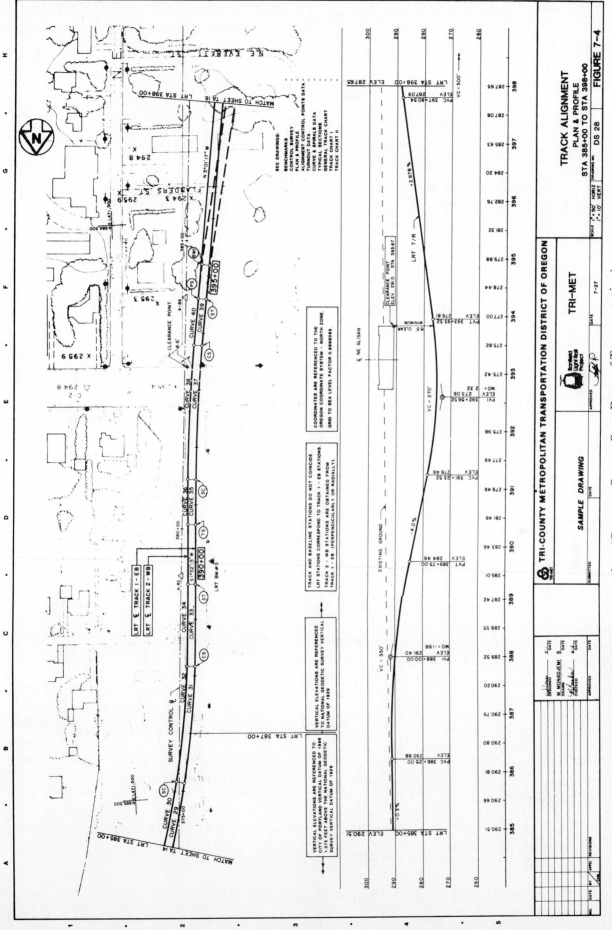

Figure 8-6 Plan and profile used in highway construction. (Courtesy Oregon State Dept. of Transportation)

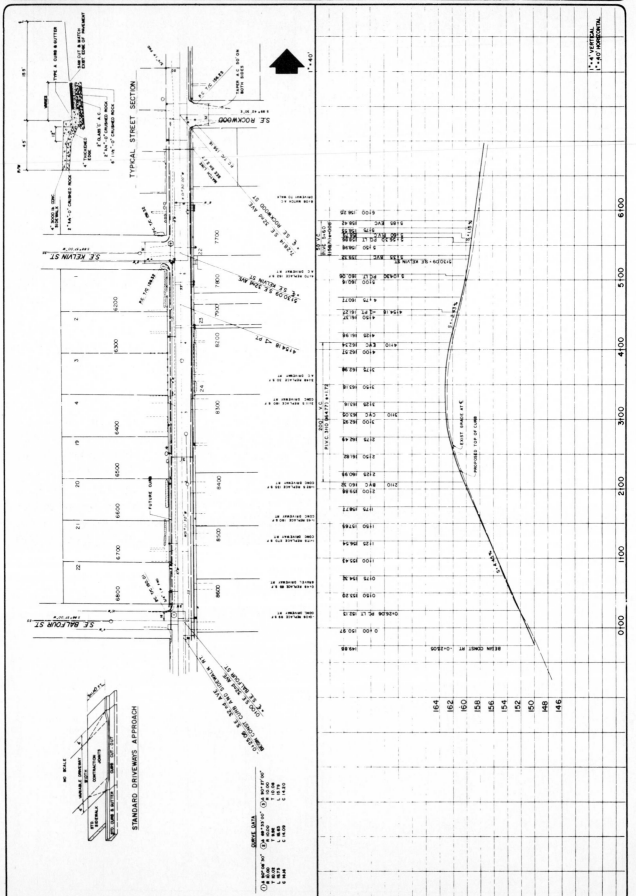

Figure 8-7 Typical plan and profile used by civil engineering companies. (Courtesy OTAK & Associates, Inc.)

so on. The example in Figure 8-7 shows a profile divided every 100 ft by vertical grid lines. At intervals along the plan view, manholes are located and their specific location is given as a station point value. The manhole symbol and station point value is shown in Figure 8-8 as they appear in both plan and profile. Manholes are abbreviated M.H., followed by an assigned number.

An important number called the *invert elevation* is always found on the profile view and is abbreviated I.E. This number represents the bottom inside of the pipe and is shown in Figure 8-8. This value is established in the design layout of the pipe or sewer line and is important in the surveying, excavating, and construction aspects of the job. Make it a habit to check the invert elevations on your drawing with either the engineer's sketch or the surveyor's field notes.

The distance between station points and the amount of slope are often indicated in the profile just above the pipe (see Figure 8-8). The size of the pipe is given first, then the distance between manholes followed by the slope, in this case vertical drop in feet per horizontal run in feet.

Grade is a common term that means an established elevation such as road grade. This is given with the manhole number and station value in the plan view. The drafter can use this number to plot grade elevation in the profile view.

Not all profiles are located directly under the plan. This is the ideal situation. Some pipelines or sewers may take several turns through a new subdivision and the plan itself is not a linear shape. This often happens and the plan may then be located to the left of the drawing and the profile on the right. The profile always appears as a straight line or flat plane, but the plan view may show several turns in the pipe. The profile is constructed in the same manner as we discussed, but the drafter is not able to project points from one view to the next.

Figure 8-8 Profile terms and symbols.

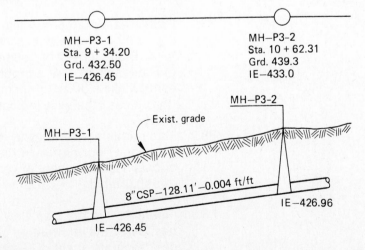

MH—P3-1
Sta. 9 + 34.20
Grd. 432.50
IE—426.45

MH—P3-2
Sta. 10 + 62.31
Grd. 439.3
IE—433.0

MH—P3-2

Exist. grade

MH—P3-1

8"CSP—128.11'—0.004 ft/ft

IE—426.96

IE—426.45

TEST

8.1. What is a map profile?

8.2. What are its uses?

8.3. Why is the vertical scale exaggerated?

8.4. Which points are projected from the map to the profile?

8.5. Who uses the plan and profile and for what purpose?

8.6. What scales are common for the plan and profile?

8.7. A point established every 100 ft in a linear survey is called a_____.

8.8. What is invert elevation?

8.9. What information may be found by a manhole symbol on a plan and profile?

8.10. What is an established elevation called?_____

PROBLEMS

P8.1. Construct a map profile along line *AB* in the map shown in Figure P7-3. The vertical scale should be exaggerated. Use an A- or B-size sheet of vellum. Label vertical elevations and indicate scales. Use a proper sectioning symbol to indicate earth. Make a diazo print and submit to instructor for evaluation unless otherwise indicated.

P8.2. This problem enables you to plot a profile of a curved road. Construct your profile in the space provided on Figure P8-2 or on a separate sheet of vellum. Use the steps discussed in this chapter.

P8.3. Figure P8-3 is the plan view of a road and underground utilities. Using the information given, construct a plan and profile on a separate sheet of C-size vellum. Show all necessary in-

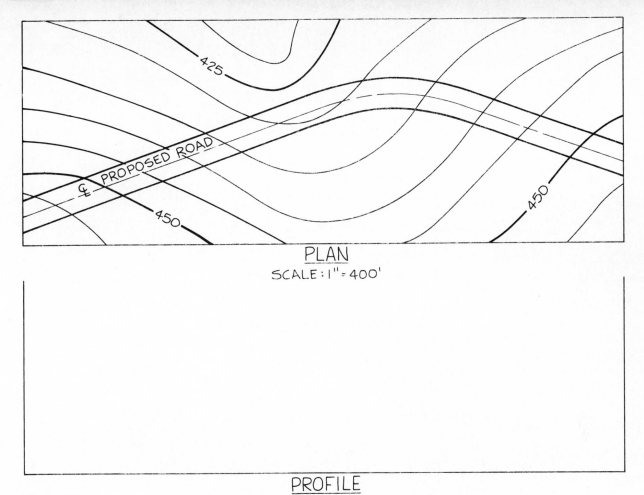

PLAN
SCALE: 1" = 400'

PROFILE

Figure P8-2

formation discussed in the text. Distances and slope between manholes should be labeled above the sewer pipe. Show water and gas lines in the plan view only. Make a diazo print and submit to instructor for evaluation unless otherwise indicated.

Figure P8-3

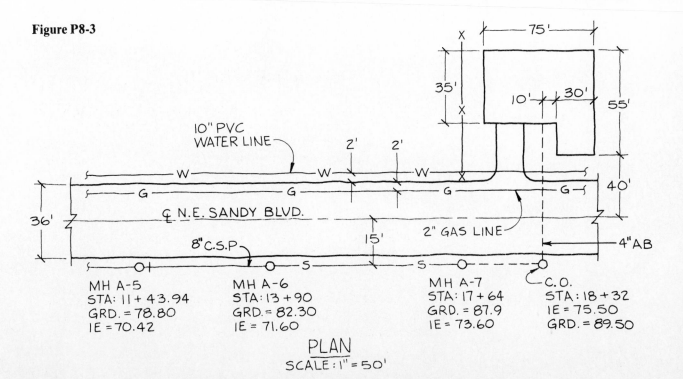

MH A-5	MH A-6	MH A-7	C.O.
STA: 11 + 43.94	STA: 13 + 90	STA: 17 + 64	STA: 18 + 32
GRD. = 78.80	GRD. = 82.30	GRD. = 87.9	IE = 75.50
IE = 70.42	IE = 71.60	IE = 73.60	GRD. = 89.50

PLAN
SCALE: 1" = 50'

9

Highway Layout

The layout of a proposed highway or road usually begins on a contour map or aerial photograph. Road designers and engineers, with input from government officials, determine the location of the road using maps and information gathered from field study. The engineer's initial design and centerline location is given to the surveyors, who then locate the proposed road's centerline and right-of-way boundaries. The surveyor's field notes are then plotted by the drafter on a contour map. When the initial layout is completed, the construction details can begin. This chapter discusses the initial layout of roads and highways.

PLAN LAYOUT Centerline (Route) Survey

Before the drafter can begin actual drawing of the road, a survey crew must physically locate the centerline on the ground and record bearings, distances, and station points as field notes. These notes are used by the drafter to plot the highway. The initial function of the survey crew is to mark the centerline of the highway. Subsequent surveys establish right-of-way, actual road widths, and other details.

Plan Layout

The plan and profile is an important drawing in highway layout. Using the surveyor's field notes, the drafter first constructs the plan view. This can be done on an existing map or the drafter may have to draw a new one. The plan view shows trees, fences, buildings, other roads and cultivated areas. The centerline or *transit line* is then drawn using bearings and distances. Station points are located every 100 ft. Figure 9-1 shows a plan layout. Notice that the road is to have curves. The information needed to construct the curve is given under the heading ''Curve Data.'' Let us examine this information.

Point of curve (PC) is the point at which the curve begins. The station value is also written at the point of curve.

Radius (R) is a curve radius. To find the center point of this radius, you must project a line perpendicular to the centerline and measure the required distance on this line (see Figure 9-2).

Delta angle (Δ) is the central or included angle of the curve. From the perpendicular line projected to find the radius center, measure the delta angle. This point is the end of the curve (see Figure 9-2).

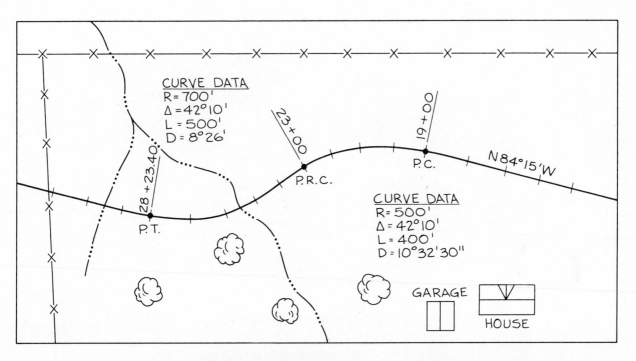

Figure 9-1 Plan layout

Figure 9-2 Basic curve data for highway layout.

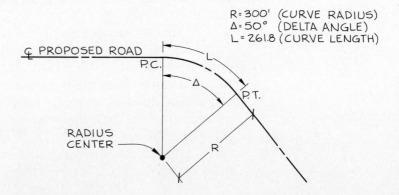

Curve length (L) is the centerline length of the curve from the P.C. to the P.T. or end of the curve (see Figure 9-2).

Degree of curve (D) is the angle of the 100-ft chord that connects station points. A chord is a straight line that connects points on an arc or circle. The degree is measured from the previous chord (see Figure 9-3).

Point of reverse curve (P.R.C.) is the point at which one curve ends and the next curve begins. A *reverse curve* is known as an "S" curve in racing terminology and contains no straight section between curves. The station value is given after "P.R.C." (see Figure 9-1).

Point of tangency (P.T.) is the end of the curve. The station value is also written at this point.

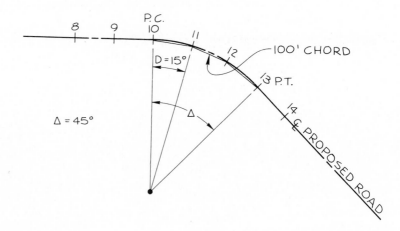

Figure 9-3   Degree of curve on a 100-ft chord.

Curve Layout: A Second Method

The survey crew may only establish centerlines and *points of intersection* (P.I.). A point of intersection on a plan view is the junction of two centerlines of different bearings. Curves are constructed at P.I.'s. Figure 9-4 shows the curve layout method using bearings and P.I.'s. The drafter is given bearings and distances and also suggested curve radii, and using this information must plot the road centerline and curves.

To find the center point of the radius, measure perpendicularly from each leg of the road to the inside of the curve, the exact radius distance, as shown in Figure 9-5. Then draw a line parallel to each leg through the radius point just measured. The intersection of these two parallel lines is the center point for the radius curve. You can now construct the curve and road outline using the radius point.

Once the centerlines of the road have been plotted and the curves are laid out, the width of the road and right-of-way can be easily drawn by measuring from the centerline.

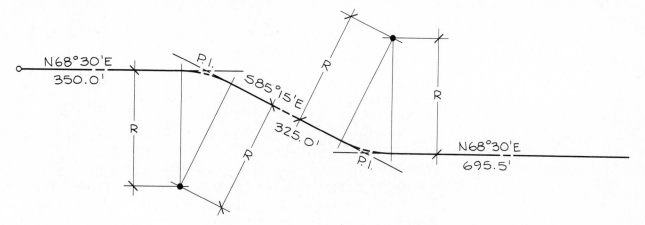

Figure 9-4 Highway curve layout using bearings, distances, and curve radii.

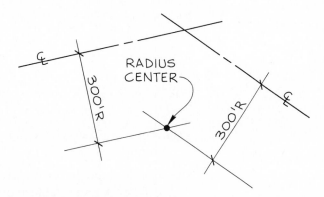

Figure 9-5 Method of locating highway curve radius center.

PROFILE LAYOUT The plan view shows all necessary horizontal control. Bearings, distances, radii, and angles. But as in the location and construction of a sewer line, a road requires vertical control. The ground does not remain flat and the road must reflect this. All necessary vertical information is shown on the profile. The layout of the profile is done in the manner discussed in Chapter eight.

Construct a grid for the profile and label vertical elevations along the side. If there are contour lines on the plan, the labels should coincide with them. Along the bottom of the profile, the stations should be labeled (every 100 ft). Next, the profile of the present ground level should be plotted, and drawn in freehand.

Vertical Curves

A vertical curve is the shape of the road or highway as it crests a hill or reaches the bottom of a valley and creates a "sag." These features are calculated mathematically by the engineer or computer program. The drafter is given all necessary elevation points and then plots the curve. In Figure 9-6, the B.V.C. (begin vertical curve) is labeled as is

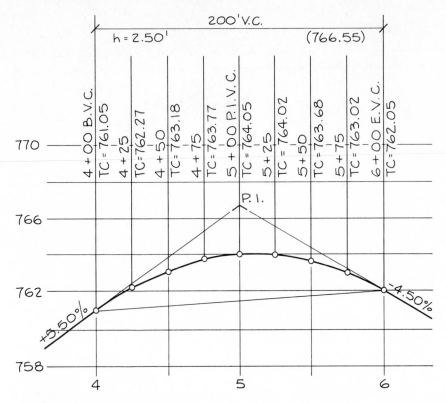

Figure 9-6 Vertical curve layout.

the E.V.C. (end vertical curve). The vertical curve occupies this entire distance and is labeled on the profile as 200 ft. V.C.

Keep in mind that vertical curves are measured as a horizontal distance and not a radius.

The P.I. (point of intersection) on the profile is the intersection of the projected grade lines or grade slopes. The elevation of this point can be calculated using the elevations of the B.V.C. and E.V.C. and grade slopes. Remember that grades are given as percentages.

The vertical curve is tangent to two points, stations 4 + 00 and 6 + 00. A straight line connecting these two stations has an elevation of 761.55 ft directly below the P.I., and the vertical distance between them is 2.50 ft. The vertical curve will pass midway through this distance and have an elevation of 764.05 ft below the P.I. The distances from the grade line to the curve at each station point can be calculated using the formula

$$\frac{(D_1)^2 h}{(D)^2} = \text{TD}$$

The tangent distance (TD) is the measurement from the grade line to the profile of the curve at a station point. Once you have solved the formula for this distance, it can be subtracted from the elevation of the grade line to give you the elevation of the road profile at a specific station point.

Figure 9-7 graphically identifies all the components of the formula. The distance between the B.V.C. and the P.I. is labeled D, and the distance from the B.V.C. to the required station is D_1. The height of the midway point directly below the P.I. is termed h. It can be stated that the distances from the grade line to the road profile are proportional to the squares of the horizontal distances from the tangent points.

Table 9-1 provides the calculations for all the points of the vertical curve in Figure 9-6.

The drafter can plot the points of the vertical curve at each station and then connect those points using an irregular curve or flexible curve. This line becomes the profile of the proposed road. Elevations of the road are written vertically at each station point.

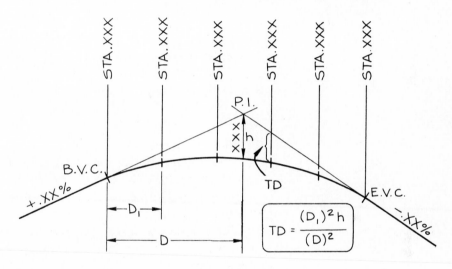

$$TD = \frac{(D_1)^2 h}{(D)^2}$$

Figure 9-7 Components of vertical curve formula.

Table 9-1 Points of the Vertical Curve in Figure 9-6

Station	Tangent Elevations	Ordinate	Curve Elevations
4 + 00	761.05	0	761.05
4 + 25	762.43	0.16	762.27
4 + 50	763.81	0.63	763.18
4 + 75	765.18	1.41	763.77
5 + 00	766.55	2.50	764.05
5 + 25	765.43	1.41	764.02
5 + 50	764.31	0.63	763.68
5 + 75	763.18	0.16	763.02
6 + 00	762.05	0	762.05

TEST

9.1. What type of survey is used to lay out the road initially? _____

9.2. What is the point at which the curve begins?_____

9.3. What is the delta angle?_____

9.4. How is the degree of curve measured? Show your answer in a sketch.

9.5. What is the point at the end of the curve called? _____

9.6. Define the point of intersection (P.I.).

9.7. Briefly describe how you would locate the center point of a 400-ft-radius horizontal curve, given two centerlines and a P.I.

9.8. Why is a profile necessary for road construction?

9.9. What is a vertical curve?

9.10. What type of information does the vertical curve show?

PROBLEMS

P9.1. For this problem you will lay out a 40-ft-wide road using the information given in Figure P9-1. Plot the centerline bearings on Figure P9-1, then construct the curves and road outlines. Label all points, bearings, and radius curves.

P9.2. This exercise involves the layout of a road using some different information. A 40-ft-wide street is to be plotted using the information given in Figure P9-2 on page 134. The street begins at point A, which is 240 ft east along Holgate Avenue.

P9.3. Get your calculators ready for this one. Figure P9-3 (see page 134) contains enough information for you to calculate the elevation of the road at each station point on the vertical curve. Using the formula given in this chapter, make your calculations and record them in the space provided.

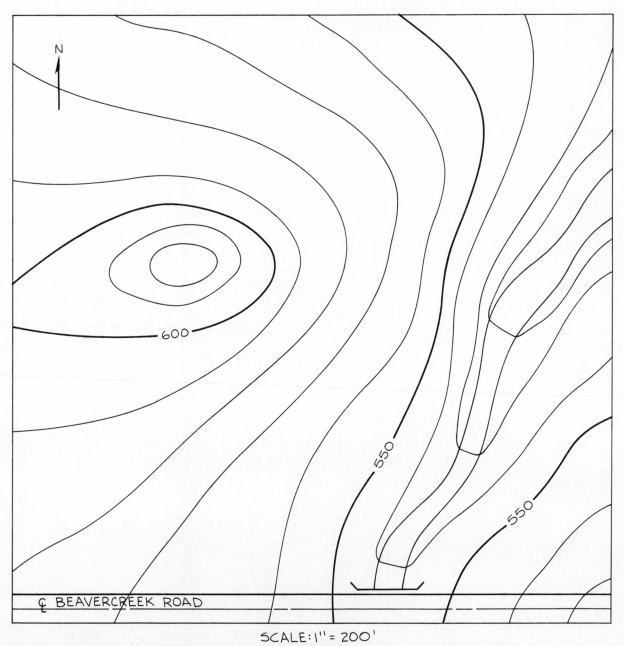

SCALE: 1" = 200'

Point A: 710' west from right edge of map on ℄ of Beavercreek Rd.
Point B: N24°30'E, 550', 200' radius curve
Point C: N39°25'W, 620', 300' radius curve, then due north to edge of map.
Road to be 40' wide.

Figure P9-1

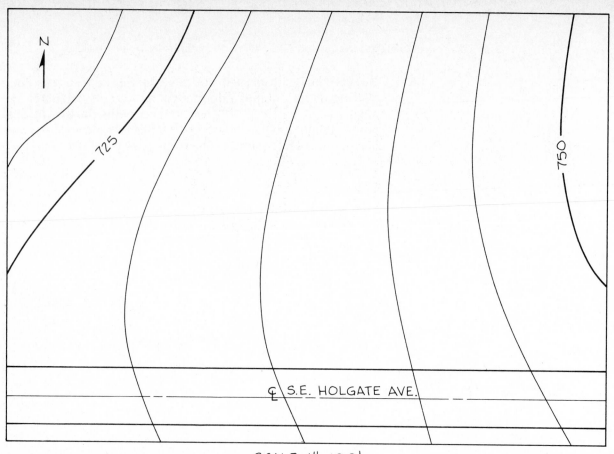

SCALE: 1" = 100'

Point A: 152' east from left edge of map on ℄ S.E. Holgate Ave.
Point B: Due north 140'
Point C: Δ angle — 30° (NE)
 Radius curve — 150' to point C

Point D: 170' from C
Point E: Δ angle — 60° (NE)
 Radius curve — 100' to point E
Point F: 160' from E

Point G: Δ angle — 90°
 Radius curve — 120' to point G
Point H: Due south to ℄ Holgate Ave.

Plot A 40' wide street.

Figure P9-2

Figure P9-3

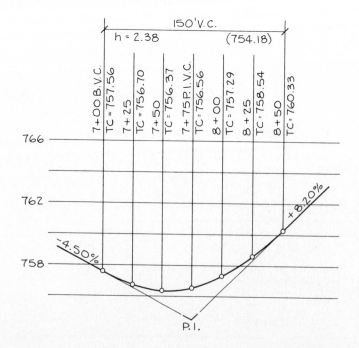

Station	Tangent distance
7 + 25	
7 + 50	
7 + 75	
8 + 00	
8 + 25	

Insert answers here

10

Cut and Fill

Cut and fill are terms used in road construction to describe the quantities of earth cut from hillsides and filled into valleys and low spots. In highway design and construction, an established grade line plotted on the profile drawing vividly shows the amount of cut and fill to be encountered. The cut and fill can be shown on the profile or on a plan view drawn specifically for that purpose. Volumes of cut and fill can then be calculated. Layout and excavation for industrial or commercial building sites also involves the use of cut-and-fill drawings.

HIGHWAY CUT-AND-FILL LAYOUT

Level Road

In designing or planning the location of a road, it is necessary to determine accurately how far the cuts and fills will extend beyond the sides of the road. This is important in determining the proper amount of land to be purchased for the right-of-way.

The road is first plotted on a contour map (Figure 10-1a). The engineer establishes an *angle of repose,* which is the slope of the cut and fill from the road. The angle of repose is basically the ratio of run to rise (Figure 10-2) and is determined primarily by the type of soil or rock to be cut through or used as fill material.

A cross-sectional view, or profile, is established off the end of the road. From the edges of the road, the angle of repose is plotted and a vertical scale identical to the horizontal scale of the map is used to mark the contour values (see Figure 10-1b).

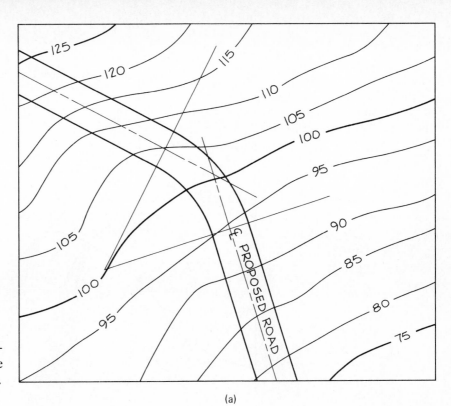

Figure 10-1(a) Begin cut-and-fill layout by plotting the new road on the map.

(a)

Figure 10-1(b) Cut-and-fill cross section is added to map perpendicular to road.

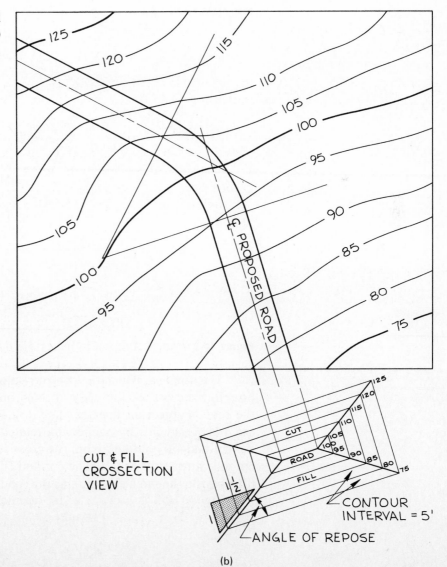

(b)

Figure 10-1(c) Intersection of cross section values with map contour lines permits outlines of cut and fill to be drawn.

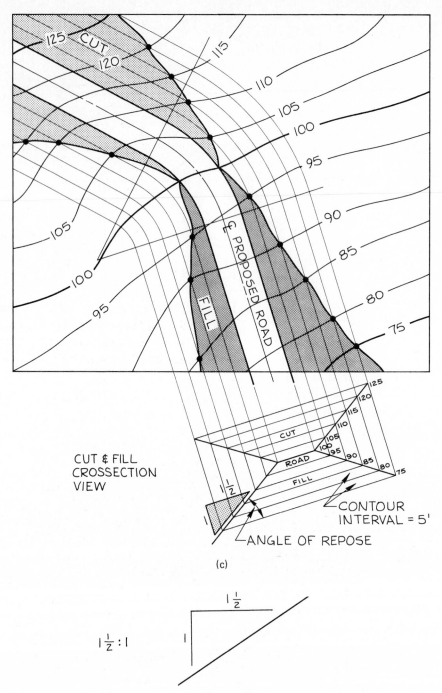

(c)

Figure 10-2 Angle of repose is the same as the ratio of run to rise.

The points at which the contour values intersect the angle of repose in the cross section are projected onto the map and parallel to the road. Project all lines as shown in Figure 10-1c. Establish the exact cuts and fills by connecting related points of elevation where the cut-and-fill contours from the cross section intersect the existing contour lines on the map. The cut and fill always begins at the contour having the same elevation as the road.

Figure 10-3 is a pictorial representation of the map in Figure 10-1.

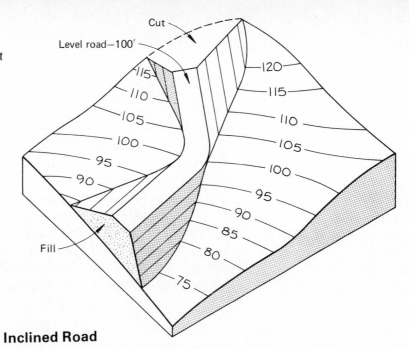

Figure 10-3 Pictorial view of cut and fill.

Inclined Road

The amount of cut and fill for an inclined road can best be established with a plan and profile. The road is plotted on the contour map as for a level road, but a profile of the existing grade along the entire centerline is drawn as shown in Figure 10-4.

Figure 10-4 Cut-and-fill layout for an inclined road.

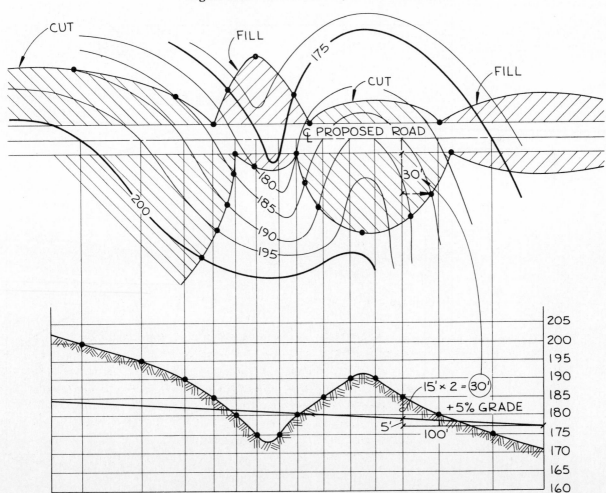

The *grade* of the road is determined and plotted on the profile. The grade is termed "percent of grade." A 1 percent grade rises 1 ft vertically for every 100 ft of horizontal distance. Therefore, a 100 percent grade is a 45° angle (see Figure 10-4).

The road in Figure 10-4 has a 4 percent grade and an angle of repose of 2:1 for both cut and fill. At the point where each contour line is projected through the profile, measure the distance from the road grade to the ground elevation. Multiply this figure by the angle of repose (2) and place that measurement perpendicular to the edge of the road in the plan. Project the point parallel to the road to the appropriate contour, as shown in Figure 10-4. Do this for all the contours that cross the road in the profile. The points can then be connected to delineate the areas of cut and fill.

SITE PLAN CUT-AND-FILL LAYOUT

Site Plan Layout

The property boundaries for the proposed site are plotted on a contour map using surveyor's notes. The elevation of the site (after excavation or filling) is determined by the engineer, as is the required angle of repose. This information may be given on the plan as well as bearings and distances for the property lines. Figure 10-5a depicts the site plan layout before application of cut and fill.

Figure 10-5 (a) Site plan layout before application of cut and fill.

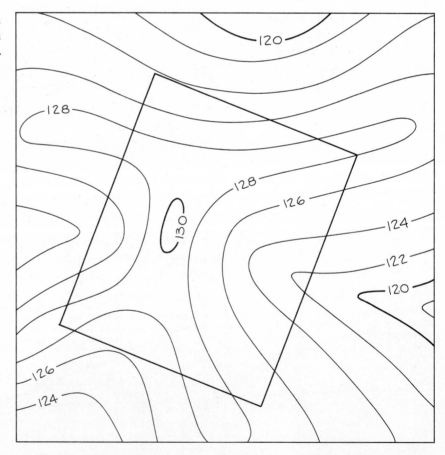

(a)

Cut and Fill

The drafter should first determine the areas of cut and fill. Then the appropriate scales for each can be plotted on the map. In Figure 10-5, the angle of repose for cuts is 2:1 and 1½:1 for fills. The contour interval is 2 ft. In areas of fill, measure perpendicular to the property lines 1½ times the contour interval and draw a line parallel to the property line. For the areas of cut, measure perpendicular to the property lines two times the contour interval, or 4 ft, and drawn lines parallel to the site boundaries. Study the example in Figure 10-5b closely.

The elevation for the site is to be 126 ft. Areas below this elevation are fill and areas above are cut. Where the first line of the fill scale intersects the 124-ft contour, make a mark. The second line of the scale is the 122-ft contour. Find all the intersections of like contours and then connect those points as shown in Figure 10-5c. Use the same technique when finding the required points for the cut. The first line parallel to the property line on the cut scale is 128 ft.

From this map the amounts of excavation and fill can be determined. Any profiles that are needed can be taken directly from the map. Legends and additional information can be placed on the map as required.

Figure 10-5(b)
Site plan layout with cut and fill scales added.

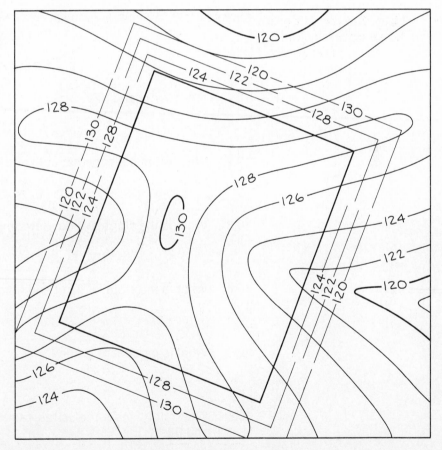

(b)

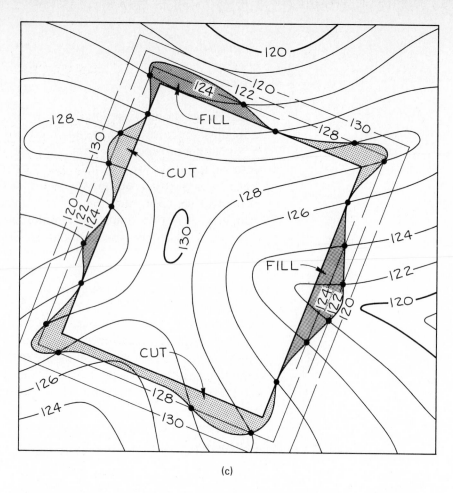

Figure 10-5 (c)   Completed site plan layout showing areas of cut and fill.

TEST

10.1.   What is cut and fill?

10.2.   Why is it necessary that cuts and fills be accurate?

10.3.   What is the slope of a cut determined by?

10.4.   What is the slope of a cut and fill called?

10.5.   What is percent of grade?

10.6.   Why might industrial sites require the calculation of cuts and fills?

PROBLEMS

P10.1. Figure P10-1 shows a proposed road layout. It will be a level road at an elevation of 270 ft. Angle of repose for cut and fill is 2:1. Construct the required cut and fill directly on the map. Your cut-and-fill scale may be placed in the space provided. Show all of your work. Use different shading techniques for cut and fill.

Figure P10-1

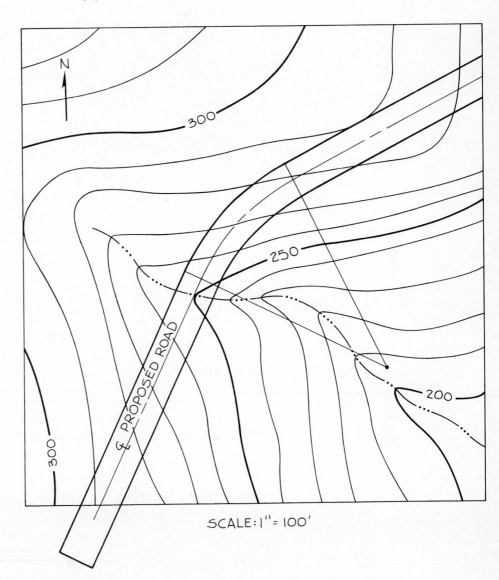

SCALE: 1" = 100'

P10.2. This exercise requires you to plot the cut and fill for an inclined road. You are given the plan, Figure P10-2, and must construct a profile from which to calculate cut and fill. Refer to Figure 10-4 if you encounter problems. The road is to have a 6 percent grade and the cut and fill begins at elevation 1500 ft. The angle of repose for both cut and fill is 2:1. Use shading and labels to identify the cut and fill in the plan view.

Figure P10-2

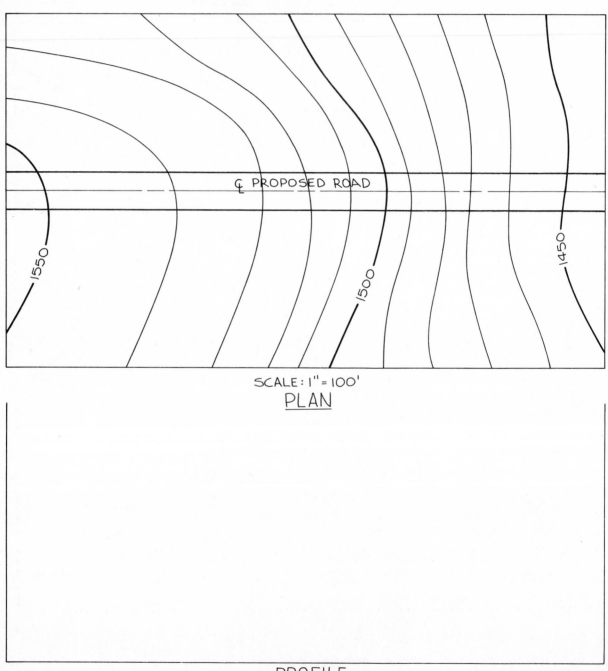

SCALE: 1" = 100'

PLAN

PROFILE

P10.3.   The cut and fill for a proposed industrial site must be determined in this problem. The site is shown in Figure P10-3. The elevation of the site is to be 640 ft. Angle of repose for cut is 2:1 and 1½:1 for fill. Plot the areas of cut and fill directly on the map and label them.

Figure P10-3

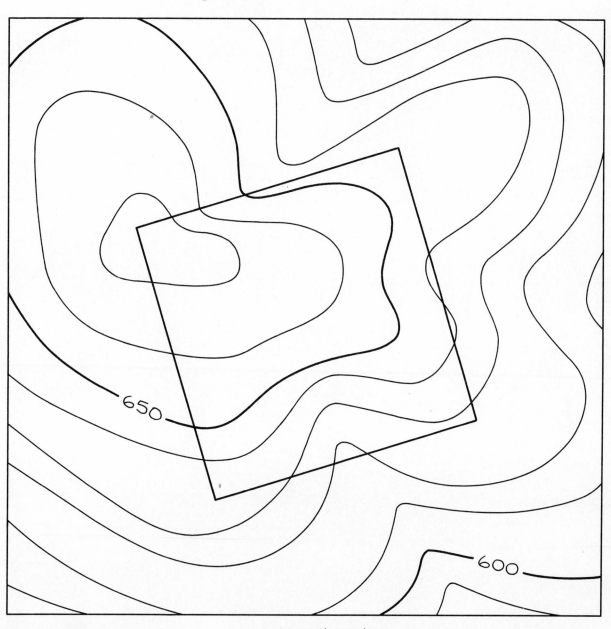

SCALE: 1"= 100'
CONTOUR INTERVAL - 10'

Glossary

AERONAUTICAL CHARTS special maps used as an aid to air travel.

ANGLE OF REPOSE the slope of cut and fill from the road expressed in feet of horizontal run to feet of vertical rise.

AZIMUTH a horizontal direction measured in degrees from 0 to 360. Usually measured from north.

BASE LINE a principal parallel used in establishing the rectangular system of land description.

BEARING the direction of a line with respect to the quadrants of a compass, starting from north or south.

BENCH MARK a reference or datum point from which surveys can begin. These are often brass caps mounted in concrete. Temporary bench marks may be iron pipes or spikes used for small surveys.

CADASTRAL MAPS large scale maps depicting features in a city or town.

CARTOGRAPHY the art of making maps and charts.

CESSPOOL a holding tank used to break down and distribute waste materials to an area of earth.

CHAIN a measurement tool composed of links originally 66 feet in length. Steel tapes of 100 feet long are often referred to as chains.

CIVIL ENGINEERING a discipline concerned with the planning of bridges, roads, dams, canals, pipelines, and various municipal projects.

CLOSED TRAVERSE a survey in which the lines close on the point of beginning or on another control point. Most often used for land surveys.

CONNECTING TRAVERSE a survey which closes on a datum or control point other than the point of beginning.

CONSTRUCTION SURVEY a localized survey in which building lines, elevations of fills, excavations, foundations, and floors are established and checked.

CONTOUR INTERVAL the vertical elevation difference between contour lines.

CONTOUR LINE a line used to connect points of equal elevation.

CONTROL POINT SURVEY the surveyor determines the elevation of important points in a plot of land. These points are plotted on a map and contour lines can then be drawn to connect them.

CROSS SECTION a profile or section cut through the land to show the shape or relief of the ground.

CURVE LENGTH the length of a highway curve from beginning to end measured along the centerline.

CUT AND FILL a road construction term that describes the quantities of earth removed from hillsides and filled into low spots.

DEFLECTION ANGLE in surveying an angle that veers to the right or left of a straight line, often the centerline of a highway, power line, etc.

DEGREE OF CURVE the angle of a chord (from the preceding one) that connects station points along the centerline of a highway curve.

DELTA ANGLE the central or included angle of a highway curve.

DISTANCE METER a surveying instrument employing electronics (lasers, radar, etc.) to accurately measure distances. Often termed an EDM (electronic distance meter).

EFFLUENT wastewater that leaves a septic tank.

ELEVATION altitude or height above sea level.

ENGINEERING MAPS detailed maps of a construction project.

ENGINEER'S SCALE a tool used by the drafting technician to accurately measure distances on a map.

EQUATOR a line that circles the earth at 0° latitude.

FORESIGHT a rod location in surveying from which an elevation and/or location reading is taken. The foresight will become the next instrument set-up point because the survey moves in the direction of the foresight.

GEODETIC SURVEY   large areas of land are mapped and the curvature of the earth becomes a factor. The process of triangulation is used.

GEOGRAPHICAL MAPS   small scale maps depicting large areas on the earth.

GRADE   an established elevation of the ground or of a road surface.

GRAPHIC SCALE   a scale resembling a small ruler in the legend or margin of the map.

GRID SURVEY   a plot of land is divided into a grid and elevations are established at each grid intersection. Contour maps can be drawn from the grid survey field notes.

HYDROLOGIC MAP   a map depicting boundaries of major river basins.

INTERIOR ANGLE   the angle between two sides of a closed or loop traverse measured inside the traverse. Also known as an included angle.

INTERPOLATION   to insert missing values between numbers that are given; an educated guess.

INVERT ELEVATION   the bottom inside elevation of a pipe.

LAND SURVEY   a survey that locates property corners and boundary lines; usually a closed traverse.

LATITUDE   an angle measured from the point at the center of the earth. Imaginary lines that run parallel around the earth, east-west.

LEGEND   an area on the map that provides general information such as scale, title, and special symbols.

LEVEL   a surveying instrument used to measure and transfer elevations. Occasionally used for distance measurements.

LOCAL ATTRACTION   any local influence that causes the magnetic needle to deflect away from the magnetic meridian.

LONGITUDE   imaginary lines that connect the north and south poles.

LOT AND BLOCK   a method that describes land by referring to a recorded plat, lot number, county, and state.

MAGNETIC DECLINATION   the horizontal angle between the magnetic meridian and the true meridian.

MAGNETIC MERIDIAN   the meridian indicated by the needle of a magnetic compass.

MAPS   graphic representations of part of or the entire earth's surface drawn to scale on a plane surface.

MAP SCALE   aid in estimating distances.

MERIDIANS   lines of longitude.

METES AND BOUNDS   a method of describing and locating property by measurements from a known starting point.

MILITARY MAPS any map with information of military importance.

MYLAR plastic media used as a base for drawings in the drafting industry.

NAUTICAL CHARTS special maps used as an aid to navigators.

NUMERICAL SCALE the proportion between the length of a line on a map and the corresponding length on the earth's surface.

OPEN TRAVERSE a survey that does not return to the point of beginning and does not have to end on a control point.

PHOTOGRAMMETRIC SURVEY aerial photographs taken in several overlapping flights. A photogrammetric survey becomes the "field notes" from which maps can be created.

PLAN AND PROFILE a drawing composed of a plan view and profile view (usually located directly below the plan). This type of drawing is often created for projects such as highways, sewer and water lines, street improvements, etc.

PLAT a map of a piece of land.

PLOT PLAN similar to a plat but showing all buildings, roads, and utilities.

PLUMB BOB a pointed weight with a line attached to the top used in locating surveying instruments directly over a point or station. In chaining, used to locate exact distance measurements directly over a station point.

POINT OF CURVE the point at which a highway curve begins.

POLYGON a multisided figure. If the included angles equal 360°, the polygon will close.

PRINCIPAL MERIDIAN is a meridian established as a basis for establishing a reference line for the origin of the rectangular system.

PROFILE an outline of a cross section of the earth.

QUADRANT the compass circle is divided into four 90° quadrants: northeast, northwest, southeast, and southwest.

RADIUS CURVE the radius (measured in feet or meters) of a highway curve.

RECTANGULAR SYSTEM a system devised by the U.S. Bureau of Land Management for describing land.

RELIEF variations in the shape and elevation of the land. Hills and valleys shown on a map constitute "local relief."

REPRESENTATIVE FRACTION distance on map/distance on earth.

ROD a square-shaped pole graduated to hundredths of a foot used to measure elevations and distances when viewed through a level or transit.

ROUTE SURVEY an open traverse used to map linear features such as highways, pipelines, and power lines. A route survey does not have to close on itself or end on a control point.

SADDLE   a low spot between two hills or mountain peaks.

SECTION   townships are divided into 36 squares—each square is a section. A section is one mile square containing 640 acres.

SEPTIC TANK   a concrete or steel tank used as a method of sewage disposal that disperses wastewater to a system of underground lines and into the earth.

STADIA   a type of distance measurement employing a level and a rod. Also a Greek term referring to a unit of length equal to 606 feet, 9 inches.

STATION   arbitrary points established in a survey usually located 100 feet apart. An instrument set-up point is often referred to as a station.

SUBDIVISION   a parcel of land divided into small plats usually used for building sites.

SURVEYOR'S COMPASS   used in mapping to calculate the direction of a line.

TERRAIN   the shape and lay of the land.

THEODOLITE   a precise surveying instrument used to measure angles, distances, and elevations.

TOPOGRAPHIC MAP   a map that represents the surface features of a region.

TOPOGRAPHIC SURVEY   a survey that locates and describes features on the land, both natural and artificial. Often accomplished through the use of aerial photography.

TOPOGRAPHY   the science of representing surface features of a region on maps and charts.

TOWNSHIPS   an arrangement of rows of blocks made up of parallels and meridians. A township is six miles square.

TRANSIT   a surveying instrument for measuring angles, elevations, and distance.

TRANSIT LINE   the centerline of a linear survey (highway, pipeline, etc.).

TRAVERSE   a series of continuous lines connecting points called traverse stations or station points. The surveyor measures the angles and distances of these lines and these field notes are then transferred to a map or plat.

TRIANGULATION   a series of intersecting triangles established as a reference in geodetic surveys. Some sides of these triangles may be hundreds of miles long and cross political boundaries.

TURNING POINT   a temporary bench mark (often a long screwdriver, stone, or anything stable) used as a pivot for a rod. The turning point can be both a backsight and a foresight for the rod.

UTILITIES   service items to a home, business, or industry such as electrical, gas, phone, or TV cable.

VELLUM   transparent paper used in the drafting industry.

VERBAL SCALE   expressed in the number of inches to the mile.

VERTICAL CURVE   the shape of a linear feature such as a road or highway (in profile) as it crests a hill or creates a sag in a valley or depression.

Abbreviations

AB	Anchor bolt	BV	Butterfly valve
ABDN	Abandon	BVC	Begin vertical curve
ABV	Above	C TO C	Center to center
AC	Asbestos cement, Asphaltic concrete	CB	Catch basin
		CCP	Concrete cylinder pipe
ACI	American Concrete Institute	CFM	Cubic feet per minute
		CFS	Cubic feet per second
ADJ	Adjacent, Adjustable	CI	Cast iron
AHR	Anchor	CISP	Cast iron soil pipe
AISC	American Institute of Steel Construction	CJ	Construction joint
		CLG	Ceiling
ANSI	American National Standards Institute	CLR	Clear, Clearance
		CMP	Corrugated metal pipe
APPROX	Approximate	CMU	Conc. masonry units
ASPH	Asphalt	CO	Cleanout
B & S	Bell and Spigot	COL	Column
BETW	Between	CONC	Concrete
BKGD	Background	CONN	Connection
BL	Base line	CONT	Continue, Continuous
BLDG	Building	CTR	Center
BLT	Bolt	CYL	Cylinder
BLW	Below	CONST	Construction
BM	Beam, Bench mark	D	Degree of curve
BOT	Bottom	D or DR	Drain
BRG	Bearing	DIA	Diameter
BSMT	Basement	DIAG	Diagonal
B/U	Built up	DIR	Direction

156

DIST	Distance	MAX	Maximum
DN	Down	MIN	Minimum
DWG	Drawing	MSL	Mean sea level
EA	Each	MH	Manhole
EDM	Electronic distance meter	MIN	Minimum
		MISC	Miscellaneous
EF	Each face	NA	Not applicable
EL or ELEV	Elevation	NTS	Not to scale
EQL SP	Equally spaced	O TO O	Out to out
EQPT	Equipment	OC	On center
EW	Each way	OF	Outside face
EXP	Expansion	OPNG	Opening
EXP JT	Expansion joint	OPP	Opposite
EXST	Existing	ORIG	Original
EXT	Exterior, Extension	OD	Outside diameter
EVC	End vertical curve	PC	Point of curve
FCO	Floor cleanout	PL	Plate, Property line
FD	Floor drain	PLG	Piling
FDN	Foundation	POB	Point of beginning
FG	Finish grade	PP	Piping
FL	Floor, Floor line, Flow	PRC	Point of reverse curve
FLL	Flow line	PRCST	Precast
FOC	Face of concrete	PRV	Pressure reducing valve
FPM	Feet per minute	PSIG	Pounds per square inch, gauge
FPS	Feet per second		
FTG	Footing	PT	Point of tangency
GPM	Gallon per minute	PVC	Polyvinyl chloride
GPS	Gallon per second	PVMT	Pavement
GR	Grade	PRESS	Pressure
GVL	Gravel	R	Radius curve
GND	Ground	RC	Reinforced concrete
GTV	Gate valve	RCP	Reinforced concrete pipe
HB	Hose bibb		
HD	Hub drain	RD	Rain drain, Roof drain
HDR	Header	REINF	Reinforce
HGT	Height	REPL	Replace
HI	Height of instrument	REQD	Required
HORIZ	Horizontal	RMV	Remove
ID	Inside diameter	RW	Right of way
IE	Invert elevation	SCHED	Schedule
IF	Inside face	SECT	Section
IN	Inch	SH	Sheet
INTR	Interior	SPEC	Specification
INVT	Invert	SQ	Square
INFL	Influent	STL	Steel
INSTL	Installation	STR	Straight
JT	Joint	STRUCT	Structure
L	Length of curve	SUBMG	Submerged
LONG	Longitudinal	SYMM	Symmetrical
LATL	Lateral	STA	Station
MATL	Material	T & B	Top and bottom

T & G	Tongue and groove	TST	Top of steel
TBM	Temporary bench mark	TW	Top of wall
TC	Top of concrete	TYP	Typical
TEMP	Temporary	VC	Vertical curve
TF	Top face	VERT	Vertical
THK	Thick	W/	With
THKNS	Thickness	W/O	Without
TO	Top of	WP	Working point
TP	Turning point	WS	Water surface,
TRANSV	Transverse		Waterstop, Welded steel

Index